▶ Eco-Cities and the Transition to Low Carbon Economies

DOI: 10.1057/9781137298768.0001

Other Palgrave Pivot titles

Paula Loscocco: **Phillis Wheatley's Miltonic Poetics**

Mark Axelrod: **Notions of the Feminine: Literary Essays from Dostoyevsky to Lacan**

John Coyne and Peter Bell: **The Role of Strategic Intelligence in Law Enforcement: Policing Transnational Organized Crime in Canada, the United Kingdom and Australia**

Niall Gildea, Helena Goodwyn, Megan Kitching and Helen Tyson (editors): **English Studies: The State of the Discipline, Past, Present and Future**

Yoel Guzansky: **The Arab Gulf States and Reform in the Middle East: Between Iran and the "Arab Spring"**

Menno Spiering: **A Cultural History of British Euroscepticism**

Matthew Hollow: **Rogue Banking: A History of Financial Fraud in Interwar Britain**

Alexandra Lewis: **Security, Clans and Tribes: Unstable Clans in Somaliland, Yemen and the Gulf of Aden**

Sandy Schumann: **How the Internet Shapes Collective Actions**

Christy M. Oslund: **Disability Services and Disability Studies in Higher Education: History, Contexts, and Social Impacts**

Erika Mansnerus: **Modelling in Public Health Research: How Mathematical Techniques Keep Us Healthy**

William Forbes and Lynn Hodgkinson: **Corporate Governance in the United Kingdom: Past, Present and Future**

Michela Magliacani: **Managing Cultural Heritage: Ecomuseums, Community Governance and Social Accountability**

Sara Hsu and Nathan Perry: **Lessons in Sustainable Development from Malaysia and Indonesia**

Ted Newell: **Five Paradigms for Education: Foundational Views and Key Issues**

Emil Souleimanov and Huseyn Aliyev: **The Individual Disengagement of Avengers, Nationalists, and Jihadists: Why Ex-Militants Choose to Abandon Violence in the North Caucasus**

Scott Austin: **Tao and Trinity: Notes on Self-Reference and the Unity of Opposites in Philosophy**

Shira Chess and Eric Newsom: **Folklore, Horror Stories, and the Slender Man: The Development of an Internet Mythology**

John Hudson, Nam Kyoung Jo and Antonia Keung: **Culture and the Politics of Welfare: Exploring Societal Values and Social Choices**

Paula Loscocco: **Phillis Wheatly's Miltonic Poetics**

Mark Axelrod: **Notions of the Feminine: Literary Essays from Dostoyevsky to Lacan**

John Coyne and Peter Bell: **The Role of Strategic Intelligence in Law Enforcement: Policing Transnational Organized Crime in Canada, the United Kingdom and Australia**

DOI: 10.1057/9781137298768.0001

palgrave▸**pivot**

Eco-Cities and the Transition to Low Carbon Economies

▶

Federico Caprotti

*Senior Lecturer in Cities and Sustainability,
King's College London, UK*

DOI: 10.1057/9781137298768.0001

First published 2015 by
PALGRAVE MACMILLAN

Palgrave Macmillan in the UK is an imprint of Macmillan Publishers Limited, registered in England, company number 785998, of Houndmills, Basingstoke, Hampshire RG21 6XS.

Palgrave Macmillan in the US is a division of St Martin's Press LLC, 175 Fifth Avenue, New York, NY 10010.

Palgrave Macmillan is the global academic imprint of the above companies and has companies and representatives throughout the world.

Palgrave® and Macmillan® are registered trademarks in the United States, the United Kingdom, Europe and other countries.

ISBN: 978–1–137–29877–5 EPUB
ISBN: 978–1–137–29876–8 PDF
ISBN: 978–1–137–29875–1 Hardback

A catalogue record for this book is available from the British Library.

A catalog record for this book is available from the Library of Congress.

www.palgrave.com/pivot

DOI: 10.1057/9781137298768

To Maria Teresa Riva and Rinaldo Caprotti

DOI: 10.1057/9781137298768.0001

Contents

DOI: 10.1057/9781137298768.0001

Preface

New-build cities are back on the agenda. Whether it is eco-cities, Smart Cities, or resilient cities, a new appetite for re-thinking the city has emerged over the past two decades. Eco-cities are one of the most popular manifestations of this trend: they bring together a high-modern interest in new urban technologies and disruptive innovations, with a concern with environmental sustainability, which has been heightened over the same time period. New-build eco-cities have been proposed in a range of diverse political and economic contexts, as responses to environmental and other issues.

The starting point for the book is a critical exploration of the notion that eco-cities are being planned, designed and built as a response to 'crisis' (broadly defined) and as a way of facing up to the impacts of various crises on an increasingly urban world. Indeed, the past two decades can be seen as a veritable 'Age of Crisis', which has seen the city beset by a range of emergencies including environmental and climate change, demographic shifts, migration, recession, real estate bubbles, energy security, terrorism, biosecurity, Peak Oil, and environmental degradation, among others. While these are real concerns in contemporary urban areas, the reactive proposal and development of urban and other policies and projects as a response to (a selection of) these crises leads to pressing questions around the direction of urban politics and the trajectory of urban futures.

This book explores the two largest eco-city projects currently under construction: the Sino-Singapore Tianjin

Eco-City, China, and Masdar City, an eco-city in Abu Dhabi, in the United Arab Emirates. Both projects are highly entrepreneurial, involving governments, global corporations, scientists, engineers and urban planners. They are also experimental, seeing new cities as bounded testing grounds for new 'ways of doing' in the city. Both eco-cities are likewise transitional, in the sense that they are conceived as engines for transition to new, more sustainable economies and urban lifestyles. In analysing these cities, the book's aim is to delve into the specifics of eco-city development: from the envisioning of new urban areas, to their marketing and presentation, to the technologies that lie at the heart of each project, and finally to the critical issues that will potentially emerge from their construction. The following, therefore, consciously takes a comparative perspective. While the social, economic, and political contexts within which Tianjin and Masdar eco-cities are being developed are very different and geographically specific, both projects are driven by environmental and economic concerns that show, at their root, remarkable similarities. Likewise, while the specific technologies and aesthetics of both eco-cities (as well as their overall scale) are dissimilar, the key drivers behind the fashioning of these new urban areas (among them the conviction that top-down planning, green building, renewable energy technologies and the market are the sources of solutions to urban and environmental problems) can also be usefully compared.

This book is therefore a brief but grounded exploration into contemporary eco-urbanism as it is expressed in eco-city mega-projects being built from scratch. One of the key issues that it repeatedly raises is the question of how – when the buildings are built, when new infrastructure is integrated into these new cities, and when the dust has settled – to ensure that new-build eco-cities are not just experiments with new technologies and with ways of enabling fluid and frictionless flows of capital, but viable attempts to include elements of *social* sustainability in the new eco-city. A view of sustainability which is concerned purely with a city's environmental footprint, or with its economic success (for investors, governments and other stakeholders) is severely limited. New eco-cities, built with technology and the market in mind but ignoring socio-economic viability, diversity and equity, run the risk of becoming mechanisms for deepening the increasingly rapid and global processes of segregation and of the worsening of inequalities. However, the book does not wish to present a bleak and gloomy picture of the eco-urban state of affairs: the focus, especially in the final chapter, is on the very

DOI: 10.1057/9781137298768.0002

real potential of proposing, finding and testing alternative forms of eco-urbanism.

I would not have been able to write this book without the support and fellowship of my family before and during the writing process. I am especially grateful to Maria Teresa Riva, Rinaldo Caprotti, Domenico Caprotti, Gabriele Caprotti, Maddalena Caprotti, Ping Gao, Yiping Wu, Rachele Riva, Emilia Rivolta, Anacleto Caprotti, and Montolivo and Gigi. I am, of course, always grateful to Eleanor Xin Gao.

Various components of the research which contributed to the development of the ideas in this book were funded by the British Academy with the Sino-British Fellowship Trust, the Nuffield Foundation, the Universities' China Committee in London, and the Royal Geographical Society-Institute of British Geographers. I am grateful for their support in funding travel and research activities. The book also benefited from Linda Berardelli's careful copyediting, from Vanesa Castán Broto's initial reading of the proposal and review of the manuscript before it went to press, and from the editorial support of Christina Brian, Amanda McGrath and Ambra Finotello at Palgrave Macmillan.

Federico Caprotti, Exeter, November 2014

List of Abbreviations

ADCO	Abu Dhabi Company for Onshore Oil Operations
ADFEC	Abu Dhabi Future Energy Company
ADNOC	Abu Dhabi National Oil Company
CSP	Concentrating Solar Power
EAD	Environment Agency – Abu Dhabi
GBES	Green Building Evaluation Standard
IPCC	UN Intergovernmental Panel on Climate Change
IRENA	International Renewable Energy Agency
KPI	Key Performance Indicator
LEED	Leadership in Energy and Environmental Design
LID	Low Impact Development
MDC	Mubadala Development Company
MEP	Chinese Ministry of Environmental Protection
MIST	Masdar Institute of Science and Technology
MOHURD	Chinese Ministry of Housing and Rural-Urban Development
PV	Photovoltaic (used mainly with reference to PV panels)
RMB	Renminbi, China's currency
SEZ	Special Economic Zone
SSTEC	Sino-Singapore Tianjin Eco-City
SSTECIDC	Sino-Singapore Tianjin Eco-City Investment and Development Corporation
TBNA	Tianjin Binhai New Area
TEDA	Tianjin Economic-Technological Development Area
UPC	Abu Dhabi Urban Planning Council

DOI: 10.1057/9781137298768.0003

1
Eco-cities in the Age of Crisis

Abstract: *Climate change, Peak Oil, energy security, and hyper-urbanisation are increasingly being identified as the 'crises' that define contemporary politics and policy. The chapter opens by investigating the 'Age of Crisis': the notion that the post-2000 period is one characterised by anxiety and constructed notions of crisis, and that these crises are focused on the city. The chapter moves on to consider the emerging trend of proposing the construction of new-build eco-cities as 'experimental cities' where solutions to multiple crises can be tested and found. The chapter closes with a critical gaze at how experimental eco-city mega-projects are being used to experiment with transition towards low-carbon economies.*

Caprotti, Federico. *Eco-Cities and the Transition to Low Carbon Economies*. Basingstoke: Palgrave Macmillan, 2015. DOI: 10.1057/9781137298768.0004.

Introduction: living in the Age of Crisis

Certain words, phrases, or images are often seen as symbols of a particular age or period in history. Often, the images associated with such periods point to the development of new viewpoints, or to technologies that are believed to represent the spirit of the age. Examples include, but are not limited to, the new artistic perspectives of the Renaissance (14th–17th centuries); engines and trains during the Age of Steam (18th–19th centuries); jet aviation in the Jet Age (1940s onwards); and Sputnik rockets and their successors during the Space Age (late 1950s onwards). The images associated with these periods are seen as symbols of the nature, character and logic of specific epochs: some of them in the distant past, some which are rather more recent. Many of these images exhibit a certain fascination with aesthetics, speed, progress, and power *over* natural and other limits.[1]

Technological developments and improvements have rarely been faster than in the early 21st century. We live in an age replete with wonder-inducing scientific and technological innovations, from advances in medicine and biotechnology, to the emergence of the internet and ubiquitous digital networks, to the first tourist space flights. Some have described the period from the 1970s onwards as the Information Age or, in more contemporary terms, as the Internet Era.[2] Others have argued that from the late 20th century on, we have been living in an as-yet undefined era *beyond* modernity, as evidenced by the proliferation of descriptors (such as post-modernity, post-modernism, post-industrial society, post-structuralism, and the like) which suggest, in the words of Anthony Giddens, 'that a preceding state of affairs is drawing to a close'.[3]

Yet, increasingly, the most widespread symbols of the current age seem not to be technological, but apocalyptic. In the mid-2010s, it seems increasingly clear that rather than living in an age wholly fascinated by technological phantasmagorias, or concerned with transition to an as-yet unspecified era after modernity, the present period can be described as the Age of Crisis. The beginnings of the Age of Crisis can be easily seen in the dashing of expectations which followed the collapse of the Berlin Wall: the 'end of history' and the heralded age of stable, liberal democracy *à l'Ouest* has not materialised.[4] Indeed, the 1990s seemed to have been characterised by unease, chaos and tragedy as much as by the development of a post-Cold War 'New Order'. In terms of war, the First Gulf War (1991), the conflict in Bosnia (1992–95), genocide in Rwanda (1994), and

DOI: 10.1057/9781137298768.0004

the war in Kosovo (1998–99), as well as many other new and continuing conflicts, underlined the stark reality that the world after the Cold War was less than stable. The World Trade Center (1993) and Oklahoma City (1995) bombings, the Falcone and Borsellino mafia murders (1992), and the assassination of Israeli political leader Yitzhak Rabin (1995) – just to mention a few events – further disintegrated the notion that with the collapse of the Soviet Union, the world would somehow become a better, safer place.

It is the events of 9/11, however, that have become the turning point for popular, media and political discourses on crisis. The terrorist attacks on New York City and Washington, DC, were an attack not only on the United States, but on its financial and political capitals: a murderous but highly symbolic thrust at the urban heart of the global financial system, and an assault on the authority of the USA as the only superpower left standing after the fall of the Wall. The cultural significance of 9/11 cannot be fully explored here, but the importance of the attacks has resonated since; in addition, 9/11 has helped to make sense of the unease and turmoil that was brewing in the 1990s.[5] The collapse of the Twin Towers focused the Global North's attention on the notion of political-ideological and socio-cultural 'crisis,' and on the 'need' to develop techno-scientific and policy-based responses to crisis.

Most closely linked to 9/11 are, unsurprisingly, discourses which have emerged proclaiming a crisis of security: this has generated and legitimised the adoption of new and more intrusive technologies of surveillance for use in the War on Terror; the militarisation of spaces of mobility, especially airports; the rapid construction of new extra-territorial systems for the projection of state power and control, such as the Guantanamo prison camp, the international outsourcing of torture, and rendition flights. Furthermore, the sense of insecurity which has pervaded much political discourse (at least in the West) since 9/11 has also raised the spectres of menaces which Western powers, used to facing up to Cold War threats, are ill-suited to tackle.[6] These include new and asymmetrical threats such as biological and chemical weapons; home-grown terrorism, as seen in the 2005 London bombings; and cyber-war, as exemplified by the massive cyber-attack on Estonia during the country's 2007 election.

The Age of Crisis has undoubtedly played into the hands of non-democratic and opaque regimes, such as China and Russia, by enabling them to enact policies and restrictive measures aimed at reducing the

DOI: 10.1057/9781137298768.0004

risk of domestic 'terrorism'. The proliferation of crisis discourses has also enabled the achievement of political, military, and industrial agendas rooted in the industrialisation of fear and responses to it: from the technological leaps and bounds involved in pilotless drone technology, to cyber-espionage, to ready justification for conflicts from Iraq to Afghanistan. This industrialisation of fear and prejudice extends to a legalised suspicion of domestic, alien or non-resident 'Others': from Latino immigrants in US cities, to construction workers from China's north-western provinces in the country's urbanising seaboard, to the furore around immigrants and non-EU students in the UK. The years since 9/11 have seen a widespread and still-simmering climate of global insecurity and fear.

One of the common characteristics of the 'crises', which have occupied centre stage since 2001, is their largely diffuse and systemic nature. The Age of Crisis is a fearful age, where crisis is not easily defined or contained within geographically delimitated space(s). Rather, it cuts across borders and affects the global arena in unpredictable ways. In this sense, Ulrich Beck's post-Chernobyl 'Risk Society' has become a 'Crisis Society', where crisis is depicted as vague but ever-lurking; a pressing concern yet only resolvable through the individualisation of risk and impacts on the one hand, and through the increasing concentration of power, violence and surveillance by the state and its agencies on the other.[7]

Nonetheless, there are two other characteristics of the Age of Crisis that set it apart from previous times of insecurity. Firstly, the crises that have been identified since 9/11 are largely focused on the city. Indeed, the events of 9/11 placed the city – as the threatened pinnacle of society – at the centre of discourses of crisis. In some ways, there was nothing new about the city being the focal point for crisis. The hopeful decade of the 1990s had the 1989 fall of the Berlin Wall as its symbolic, urban root. The darker, less hopeful side of that initial post-Cold War decade was rooted in an equally powerful and deeply urban event: the massacre of protesters in Tiananmen Square, Beijing, in the same year. After 9/11, the crises of terrorism, biosecurity, SARS, finance, the subprime mortgage crisis, urban unrest from London and Paris to the Arab Spring, the 'problem' of demographic change and rural-urban migration, and other crises, have been largely urban in character. The city is the stage on which the Age of Crisis is playing out, and it is one of the places where crisis becomes most visible and concentrated.

DOI: 10.1057/9781137298768.0004

Secondly, one of the major trends of the Age of Crisis is the focus on *environmental* crisis. This has roots in the environmental movement from the 1960s onwards, and also has deeper foundations in earlier unease about the alienation of humanity from 'nature'.[8] However, environmental crisis had become wholly accepted and institutionalised as a leading discourse from the 1990s onwards. Key turning points were events such as the 1992 Rio Summit; the formation of the UN's Intergovernmental Panel on Climate Change (IPCC) in 1988 and the issuance of the first of its series of climate change reports from 1990; and the yearly series of UN Climate Change Conferences which began in 1995.

The environmental crisis is, therefore, one of the most diffuse, cross-border, and potentially destructive crises that have gained media, political and cultural attention in the past 20 years. It has been characterised by the sense not only that the environment is being damaged through the workings of our current oil-addicted, capitalist consumer society, but also by an explicit assumption that the environmental crisis that we are now facing is, in some ways, *the* crisis to end all crises. This is because climate change, rising sea levels, and the effects thereof are constructed as somehow (and perhaps appropriately) millenarian and apocalyptic, being deeply rooted in the changing character of late modern society.[9] At the same time, environmental crisis is often depicted as deeply urban: Peak Oil scenarios are seen as having the greatest impact on urban agglomerations; rising sea levels threaten to swamp cities; the increased frequency and severity of climate hazards such as storm surges and floods place the city, and with it modern society, directly in the firing line. Indeed, the IPCC's 2014 assessment of climate change impacts will, for the first time, include an evaluation of the changing climate's effect on cities.[10]

The link between environmental crisis and the city has been underlined to a great extent by some of the great urban environmental disasters which have occurred since 9/11: Hurricane Katrina's destruction of New Orleans (2005), the tsunami and associated nuclear disaster at Fukushima, Japan (2011), and typhoon Haiyan's destruction of Tacloban city in the Philippines in November 2013, among others. At the same time, cities (especially those in emerging economies such as China and India) are seen as straining under the environmental burden of increasing levels of consumption, waste and contamination. The prospect of 'climate refugees' flooding into well-protected, wealthy urban areas, and of millions of displaced and dispossessed urban dwellers from coastal

DOI: 10.1057/9781137298768.0004

areas in the Global South suffering from overwhelmed and unequally distributed infrastructures, as well as the depiction of the city as a focus for perilous outbreaks of new and existing diseases (from cholera to H5N1) as a result of climate change, paint a picture of an urban world in peril as the result of a developing environmental crisis.

These twin aspects of the Age of Crisis – the city as the locus of crisis, and environmental crisis – have therefore become largely linked. The environmental crisis, diffuse and trans-border in character, is most readily identifiable in its impacts on urban areas. This is in part reflected in the resultingly wide range of recent depictions of the environmental destruction of the city in popular culture. From the 2004 movie *The Day After Tomorrow*, in which cities such as New York are shown being encased in ice as a result of hyper-rapid climate change, to Al Gore's *An Inconvenient Truth* (2006), a film which highlighted the trends converging towards climate crisis, to the 2009 on-screen adaptation of McCarthy's bleak *The Road*, in which a nuclear war-induced environmental collapse has destroyed cities and left human society in a depleted downward drift, the city and the environment are increasingly seen as locked into a negative spiral.

One of the evocative symbols of this cultural trend towards considering the environmental crisis as an *urban* crisis is the cover of the September 2013 issue of *National Geographic*. Titled 'Our rising seas,' the magazine's main story highlighted the potential for large-scale damage to cities as a result of rising sea levels. The story's focus on 'super-floods' was compellingly illustrated by the cover image of the Statue of Liberty partially submerged, with a line marking the level to which the water would rise if the whole cryosphere (the parts of the Earth covered by ice) melted. What the image does not show is equally important: the high-water mark on the statue implies a disastrous effect on Manhattan, which would presumably be blotted out by rising seas, with just the tops of tall buildings jutting out of the water.

The establishment of a link between environmental crisis and the city has, in turn, energised the idea that in order to confront the increasing risks associated with environmental crisis, what must be changed is the city: its physical character, the consumption and production patterns associated with urban life and urban economies, and the technological and material infrastructures which constitute the city proper. For example, in *Resilient Cities*, a recent book by urban and environmental scholars Peter Newman, Timothy Beatley and Heather Boyer, the main argument

DOI: 10.1057/9781137298768.0004

is that climate change and resource depletion, and particularly the Peak Oil scenario, are the slow-moving but continually unfolding crisis contexts that threaten contemporary cities.[11] The ability of cities to be *resilient* and to *adapt* to changing conditions and threats is highlighted as crucial to the survival of urban areas. Although *Resilient Cities* is limited in its identification of oil dependence as the source of ills such as war and inequality – many others would argue that the problem lies with the economic system that is powered by oil, and not with oil itself – it is nonetheless illuminating to consider it as an example of how the city is placed at the centre of debates on responses to environmental crisis.

Newman, Beatley and Boyer highlight four scenarios for the urban future, based on the response of cities to climate change and Peak Oil. Firstly, they identify a 'collapse' scenario, defined as a rapid decline resulting from an inability to adapt. As an example of this, they cite the 'collapse' of New Orleans in 2005 following hurricane Katrina:

> [T]he city of New Orleans collapsed due to an extreme climate event. The lack of preparation in the city was scandalous. Once all civilizing constraints disappeared people tried desperately to find food and safety. The scenes shocked us all. But history suggests that we shouldn't be shocked – the potential is there in any city.[12]

Their second scenario is the ruralisation of the city, where urban areas become less dense and more reliant on the local production of food and other resources, although this scenario is tainted by the reality that many cities will be unable to be anything close to self-supporting, and will therefore rapidly and disastrously decline before being able to respond and adapt. This 'revolutionary collapse' is at once both apocalyptic and utopian, as an idealised society grows from the ashes of a collapsed capitalist urban arena. However, as Newman, Beatley and Boyer wrote, 'Mao and Pol Pot certainly thought this way was the only morally correct lifestyle,' and one of the lessons of the 20th century has been that the potential for authoritarian and totalitarian results goes hand in hand with utopian thinking.[13]

The third scenario, which Newman, Beatley and Boyer call the 'Divided City,' is based on the idea that the wealthy will recognise the need to respond to the climate and oil crises and will construct high-tech, sustainable urban enclaves for themselves while leaving the poor on the 'fringes' of the city, reliant on ever more expensive oil and gas and unable to benefit from the expensive renewable energy sources and technologies, as well as green buildings, made available by and for the wealthy.

DOI: 10.1057/9781137298768.0004

These poor areas are prone to collapse, thus prompting the evolution of a stark dividing line between ecologically resilient, wealthy enclaves and poor, bleak outlying areas. To some extent, this is a state of affairs that has already been in existence for some time: flights into Mumbai almost invariably pass over thousands of hectares of slums suffering from socio-ecological and other inequalities, while gleaming office towers rise on the horizon.

The fourth and most amenable scenario is that of the resilient city, which is essentially an expansion of the wealthy enclaves of the 'Divided City' so that the technologies and planning interventions featured in those elite enclaves become available to all. Nevertheless, there is a distinctly utopian feel to the authors' description of the resilient city:

> People will have access to jobs and services by transit or walking as well as the use of electric cars for short car journeys. Intercity movements will move towards fast electric rail and will be reduced considerably by the new generation [of] high quality interactive video conferencing. Green building design and renewable fuels will be a part of all neighborhoods. The city will develop rail links to all parts of the city, walkable centers will be created across the city-region using the best green buildings and infrastructure... [14]

The quasi-utopian ideal of a 'resilient city', as discussed above, highlights a broad, emerging concern with facing up to the challenge of the city in crisis.[15] Millenarian and apocalyptic discourses in popular media and policy, coupled with scientific corroborations of the fact that the climate is not stable and unchanging, but open to anthropogenic influences, has inevitably led to stark conclusions about our urban future.

Eco-urbanism, transition and the eco-city

In light of the environmental and oil crises identified above, and in the context of increasingly rapid global hyper-urbanisation, eco-cities have come to the forefront of national and global agendas. This trend has included the adoption of eco-city ideals by international organisations whose remit focuses, in whole or in part, on sustainable development and sustainable urbanisation. For example, the World Bank introduced its *Eco2 Cities* urban sustainable development framework in the late 2000s, focusing on 'second generation' eco-cities as central nodes in delivering sustainable urbanism.[16] Non-governmental organisations such as Ecocity Builders, based in Oakland, California, are also active across a range of

DOI: 10.1057/9781137298768.0004

remits (from education, to international conferences, to consulting) concerned with eco-cities as solutions to pressing environmental and other issues.[17] In addition, existing national and municipal governments have been active in putting eco-city planning at the forefront of their sustainable urban development agendas. This includes eco-city development plans in Melbourne, Australia, and nation-wide plans for the construction of 13 'EcoCités' in France, amongst many other examples.[18]

In a broader sense, eco-cities can be placed within a framework of increasing focus on 'green' urban solutions, in academia and in the realms of urban policy and practice. This broad framework can be termed 'eco-urbanism,' and it has gained popularity and prevalence since the early 2000s.[19] It includes urban interventions including green master-planning, green building, sustainable urban design, and sustainable transport. The eco-city concept is where these multiple facets of eco-urbanism have coalesced in recent years, and this is nowhere more evident than in the wide range of eco-cities that have been planned and designed across the world within the first 15 years of the 21st century. Eco-cities are being proposed as 'solutions' to a range of problems in the Age of Crisis, most of which (but not all) are environmental and related to humanity's effects on climate.

A key feature of many eco-cities being designed or constructed in the 21st century is their experimental nature. Indeed, the eco-city can be thought of as an *experimental* city, where new technologies, policies, and 'ways of doing' can be tested and trialled. The eco-city has been identified as a technical and 'scientific' repository and container of potential solutions to the crises mentioned above. This is exemplified most clearly in the range of technological 'solutions' that are central to contemporary new-build eco-city projects. For example, Masdar eco-city, in Abu Dhabi, is largely focused on the use of solar power to provide for the new city's energy needs, and the stalled Dongtan eco-city project, near Shanghai, prominently featured zero-energy buildings. The dominant focus on technology as a key feature of eco-cities can also be seen in the fact that most contemporary eco-city plans and marketing documents are seeded with references to renewable energy, green architecture, green space, and high-tech systems such as: the Personal Rapid Transport pods being trialled in Masdar eco-city, a municipal pneumatic waste collection system in the Sino-Singapore Tianjin Eco-City, China; and the segregation of cars from view and from pedestrians and cyclists by building Tianfu Ecological City, near Chengdu, China, on several levels.

DOI: 10.1057/9781137298768.0004

The latter exemplifies the macro view of the eco-city as an experimental whole: in the contemporary eco-city, the infrastructure and techno-economic systems which course through the city are being opened up to experimentation.

The view of eco-cities as testing grounds falls within the broader experimental remit of ecological urbanism. Indeed, many of the initiatives – in governance, infrastructure, transport, and the like – which can be labelled as 'eco-urban' are characterised by attempts to trial new approaches to the city in the context of environmental and other crises. In her book, *Cities and Climate Change*, Durham University geographer Harriet Bulkeley argues that certain urban projects and interventions can be seen as 'climate change experiments' as long as they aim to develop new rules or regulations, seek to govern the conduct and actions of others, and are designed and formed beyond traditional channels of decision-making.[20] In this sense, eco-cities fit squarely into the definition of 'climate change experiments'.

Eco-cities' experimental character is not confined to the various features and facets of urban living and its associated infrastructures, buildings, and city lifestyles. One of the key ways in which eco-cities are being seen through an experimental lens is not only as testing grounds for new technologies and policies which have effects contained within the eco-city's boundaries. Rather, many contemporary blueprints view eco-cities as experimental pivots within much larger strategic visions centred around a transition to a low-carbon, or 'green,' economic basis for a whole region or state. In this sense, eco-cities can be understood as fitting within generally state-led initiatives aiming to redirect national economies away from carbon-intensive economic bases.

The view of the city as a pivot of transition strategies is part and parcel of a wider trend towards trying to *understand* and *harness* processes of transition. Scholars from a wide range of disciplines, working in the broad field known as transition theory, have attempted to understand and identify the factors which help cause shifts in social and technical systems, and through which innovations are adopted. These may seem like abstract questions, but on an urban level they concern the ways in which cities come to be shaped in the ways we are familiar with today. For example, scholars of transition have tried to understand how technologies that are now ubiquitous in the city and in wider society (from piped water systems, to electricity) have come to be widely adopted, very often in a short space of time.[21] Studies of transition have, therefore,

DOI: 10.1057/9781137298768.0004

in large part tended to focus on how innovations arising in specific, bounded contexts then come to influence and define wider societal, political, economic and cultural contexts. Thus, transitions theorists ask the question of how innovations and disruptive technologies emerge in socio-technical 'niches', in which they are protected and stimulated, and from which they are then adopted in wider society. When wider societal adoption occurs, then a socio-technical transition is deemed to have taken place.[22]

Often, questions about how and why certain transitions happen are tackled by using a historical approach, considering transitions which have happened in the past in order to inform an understanding of the potential transition mechanisms and pathways which may take place in the future, and which may therefore be influenced in specific directions. This is key, as identifying potential transition pathways is a key part of strategic planning (at the level of cities, regions, countries and the globe) aimed at achieving a specific future state.[23] This is where the city as a key site for transition strategies really comes into its own: by acting as an experimental site that can be treated as a bounded and measurable experimental niche. Identifying eco-cities as potential transitional niches usefully moves past one of the most powerful critiques that have been made of transition theory: that theorisations of transition have been largely abstract and placeless, and that more attention needs to focused on the *places* where niches emerge.[24] A focus on eco-cities effectively spatialises the study of transition strategies that have emerged as a response to the crises mentioned above.

In part, seeing the city as central to transition is nothing new: witness the pivotal role played by cities such as Shenzhen in the course of China's embrace of capitalism from the late 1970s onwards. However, the growth of Shenzhen as a city was not, by and large, planned for. What differentiates many current eco-city experiments from cities such as Shenzhen is that new eco-cities are being envisioned and built from scratch as engines and testing grounds for transition. Several of these cities are being planned to an extraordinary level of detail – although many of those plans often seem to remain firmly on blueprints and proposal documents without materialising into new urban environments.

The view of eco-cities as transitional experiments need not, however, be confined to large-scale, macro views of the city as a whole in its role as a transitional engine for a peri-urban or wider region. There are several examples of proposals to use existing and new cities as transitional

DOI: 10.1057/9781137298768.0004

experiments focused not on 'the economy' as a whole, but on specific sectors and sub-sectors of the green economy, broadly understood. This is because focusing on a single set of transitional experiments (for example, policy trials to promote green building technologies) can be useful in not only testing the efficacy of policy for promoting transition on a smaller scale, but also as ways of calibrating techno-industrial policy interventions. This enables policymakers to track the diffusion of innovative technologies and products within existing economies and socio-technical systems, and also enables the identification of obstacles to diffusion, and assessments as to whether, in the absence of widespread diffusion, learning has nonetheless taken place which will influence the direction of a dominant socio-technical system.[25] Therefore, the adoption of a new technology or socio-technical system at the 'niche' level within a new city project may eventually have much broader effects within and outside the city, and in wider society.[26]

Eco-cities' transitional character can therefore be viewed through the lens of what has been called 'systems innovation'. This assumes that change takes place incrementally and gradually, and over considerable spans of time. However, the development of eco-cities as responses to 'crisis' is also evidence of an approach to socio-technical change that places a focus not simply on gradual societal change, but on the role of crises, or 'shocks', in enabling societies to learn to adapt and change: change can result from the learning enabled by a single crisis event, or by a series of shocks.[27] In the case of eco-cities, what is interesting is the element of forecasting which is evident in planners' blueprints for these urban projects: attempts to fashion the city so as to be more resilient to future shocks, such as energy crises, by adapting the city (partially, at least) to a series of envisioned *future* shocks. Therefore, in addition to their strategic role as engines of gradual transition, eco-cities are also transitional and experimental in their character as urban areas built as a response to crises which have not yet occurred, but which are expected to have significant impacts on the cities of the future.

Finally, contemporary eco-cities' experimental character can be seen in the detailed focus on measuring and assessing the *performance* of eco-city projects according to specific technical, economic, environmental, and (to a lesser extent at the time of writing) social indicators. Several eco-city projects involve the establishment of Key Performance Indicators (KPIs) as well as other indicator systems and ways of measuring the success and development of the city. These frameworks are, at times, developed as part

DOI: 10.1057/9781137298768.0004

of the master plan for the city or as part of wider strategic thinking about the role of the city in low-carbon transition trajectories. In other cases, eco-city projects incorporate indicator and performance frameworks, but also rely on external actors (such as consultancies) to independently assess nascent eco-cities' performance. However, there currently exists a variegated geography of indicator systems, and an equally complex landscape of motivations for the establishment of indicator and evaluation frameworks for eco-cities by local and national governments, and by international bodies.[28] Nonetheless, what is clear is that the eco-city is being envisioned as an experiment, and evaluated as such: the new urban projects being developed worldwide exist under the microscope of policymakers, corporations, consultancies, planners, and bureaucrats.

Thus, eco-cities can be seen as experimental insofar as they seek to establish new rules and regulations; aim to change citizen behaviour (even if the ways in which this will be done is often less than clear); and are designed and planned in ways unlike most other city planning projects.[29] On this last point, the lack of involvement of communities in many eco-city enterprises, and the parallel, strong collaborative links between government and the private sector, often seem to tinge eco-city projects with a less than democratic aura. However, as seen above, eco-cities can also be understood as *much more than climate change experiments*. In their focus on economic and socio-technical transition, and in their aspects of socio-economic planning, eco-cities can be understood as transitional experiments rooted in concerns about environmental crisis. Nonetheless, in different contexts the crises to which individual eco-city projects are expected to respond vary. Thus, in the case of Chinese eco-cities, environmental concerns are often married to the need to 'solve' the urban crisis stemming from rural-urban migration, and to the imperative of using the city as an engine for transition to a greener, high value economy. In the case of Abu Dhabi, the environmental and economic crises which new eco-cities such as Masdar eco-city are meant to tackle are rolled up together into concerns around the Peak Oil scenario, which is a key issue in an emirate whose economy is largely based on oil revenues.

Geographies of the eco-city

The eco-city concept has a distinct history. However, its roots are also deeply entwined with the history of urban planning from the 19th

DOI: 10.1057/9781137298768.0004

century onwards, and with the ongoing narrative of the relationship between nature and the city in modernity. As a specific concept, the idea of the eco-city as it is described today can be traced to the 1970s and to concerns with building urban areas which were not only economically and socially viable, but also ecologically sound.[30]

Much of the initial driving impetus behind the development of eco-city ideas was urban theorist and practitioner Richard Register's attempts to think about, and plan, new cities which would encapsulate the ideal of a balance between society and nature. In 1975, Register and his collaborators founded Urban Ecology, a non-profit organisation whose aim it was to reshape cities so as to make them more harmonious vis-à-vis nature and the environment. In 1987, many of Urban Ecology's initial ideas were sharpened with the publication of *EcoCity Berkeley*, Register's manifesto for how the city of Berkeley, California, could be rebuilt according to more ecologically friendly principles. The idea of linking eco-city *theory* to *practice* in actually existing urban environments became popular with academics, planners and policymakers, and in 1990, the first *International Eco-City Conference* was held in Berkeley: it was attended by over 700 scholars, urban planning practitioners, environmentalists and other interested parties. From those beginnings sprang a series of international eco-city conferences, held every two years in different locations across the globe: their character remains one of deep engagement between different actors and institutions interested in eco-cities.[31]

Throughout the 1990s, the eco-city concept continued to gain traction, as witnessed by the 1992 publication of David Engwicht's *Towards an Eco-City*, republished in the USA the following year with the title of *Reclaiming our Cities and Towns*.[32] Much of the focus of these early efforts was primarily on reshaping *existing* cities and urban areas so as to make them more environmentally friendly. Many of the initial proposed interventions by eco-city advocates focused predominantly on highly visible aspects of cities' 'negative' impacts on nature, such as congested and polluting transport networks and dependence on automobiles. In addition, in the 1990s many of the more critical 'voices' in urban studies (especially those drawing on the left critical tradition, but also including many in the environmental movement, broadly defined) did not seem to engage with the emergent set of discourses around eco-city ideas. This would have been useful in that critical scholars have much to contribute on the links between the workings of contemporary capitalism (with its multiple 'symptoms', from suburbs, to environmental despoliation,

DOI: 10.1057/9781137298768.0004

to environmental and socio-economic inequalities), and increasingly vocal proposals to think of eco-cities as providing a 'cure' to these ills.[33] In addition to a relative dearth of critical thought or scholarship on the eco-city movement in the 1990s, little serious attention was given by policymakers, planners, and most scholars, until the early 2000s, to the idea of planning and building large new cities from scratch along eco-city principles.

In terms of planning history, the development of the eco-city concept drew on a long history of attempts to craft cleaner, greener, and more amenable cities and towns from the 1800s onwards. From Ebenezer Howard's Garden Cities, to New Towns in the Soviet Union, fascist Italy and New Deal America, the late 19th century and early to mid-20th century was replete with attempts to *change* the city and to build new cities to reflect political and ideological ideals, and to materialise norma-tive views of how the city *should* look, behave, and function. Many plans for New Towns and neighbourhoods in the 20th century squarely focused on the links between technology, society, nature, and ecology, prompting historians Robert Kargon and Arthur Molella to define the wide range of new urban areas built during the 20th century as 'techno-cities'.[34]

In addition to the antecedents of eco-city discourses in the history of urban planning, the nascent eco-city movement drew on wider concerns with the environment that emerged from the 1970s onwards. More recently, these concerns have (rather opportunistically for planners and developers, who stand to gain from building new cities) dovetailed with the pressing economic, security and resource issues identified above when discussing the Age of Crisis. In other respects, however, many other trends have contributed to the placing of eco-cities at the juncture between socio-economic and environmental planning and policy. These include the development of the green movement, bioregionalism, and social ecology, as well as discourses around the desirability of sustainable development from the early 1990s onwards.[35]

Eco-cities and eco-neighbourhoods started to be built in a wide range of settings from the early 2000s, although earlier isolated examples can also be found. Initially, many early projects focused on eco-neighbour-hoods, individual green buildings, or on greening existing infrastructure, such as transport, or networks to provide energy and heat to public hous-ing. These types of projects were clearly experimental, and many of them featured a transitional purpose. An example of this can be seen in the trialling of low-carbon innovations in social housing developments in

DOI: 10.1057/9781137298768.0004

Ljubljana, Slovenia: rather than rebuilding housing from scratch, the city government introduced a series of initiatives aimed at delivering technological innovation and transition within existing residential buildings in the city.[36] Nevertheless, the years since the start of the millennium have seen the development of several different types of projects that tend to be included under the eco-city banner.

In order to come to an understanding of this wide range of projects branded as eco-cities, Simon Joss and his team at the University of Westminster's International Eco-Cities Initiative have developed a global eco-city survey and classification typology.[37] In their work, they identify three distinct types of urban developments that are generally described as eco-cities. The first kind are new-build eco-cities: whole new urban areas built from scratch along 'eco' principles and following the codes set down in an eco-city master plan or blueprint. Examples of these types of projects are the new eco-cities being constructed as part of national transitional strategies in China and the United Arab Emirates. The second category is the expansion of existing urban environments through the construction of new neighbourhoods or towns, such as some of the projects included in the now-stalled UK Eco-Towns government initiative started under Tony Blair's Labour government. For example, the planned construction of Bicester Eco-Town, in Oxfordshire, was a clear expansion of the existing urban area in Bicester, aimed at controlling the expansion of Bicester as a commuter town on the Chiltern railway line into London. The third type of eco-city development is the retro-fitting of existing urban environments, as exemplified by initiatives such as the *EcoCity Alexandria* project. Set up in 2007, the project aims to apply retro-fitting visions to facets of the urban area in the American city of Alexandria, Virginia. These facets include land use planning, open spaces, and water, among others.[38] To some extent, therefore, the spectrum of developments that can be defined as eco-cities stretches from new-builds to the more adaptive initiatives aimed at changing existing cities, neighbourhoods, and specific buildings.

To date, eco-city developments show a distinctly global geography, with some notable geographical 'gaps'. According to Joss' survey, Europe and Asia (with 69 and 70 projects respectively) lead in terms of the number of eco-cities planned or under construction, although the Asian total is significantly bolstered by the large number of Chinese eco-cities under development. With 25 projects on the books, the Americas feature much less prominently on the global eco-city landscape. However, it is Africa and the Middle East, with just 10 eco-cities planned across these macro-regions in 2011, which represent the largest geographical areas where eco-cities, or indeed eco-urbanism of any type, are not making significant inroads.[39] At national or regional scales, there is evidence of a very variegated landscape of eco-city development: Russia featured just a single project by 2011, for example, compared with seven in Sweden. Likewise, the USA and Canada account for 20 out of 25 eco-cities in the Americas: Brazil, Panama and Ecuador are the only other countries with projects under development (or, in the case of Curitiba, Brazil, cities which have already been completed). Similarly, central Asia features no eco-city developments: Eastern Europe is also bereft of them. Therefore, it can be argued that the global geography of eco-cities is highly concentrated in particular national contexts, and almost completely absent in many countries.

The uneven geographies of eco-city project locations raise some important questions. Firstly, it is crucial to identify the main drivers of eco-city development: these drivers (such as the state-led sustainable city policies, competitions and incentives which are common in China), be they largely institutional or market-based (or a blend of the two, as in the case of Masdar eco-city, Abu Dhabi), provide the background context through which the motivations and stated aims of eco-cities can be analysed and interrogated. Secondly, if eco-cities are being proposed as solutions to the concerns raised against the background of the Age of Crisis, then these projects need to be investigated in light of this context. For example, it is increasingly clear that the effects of climate change are likely to impact most negatively on coastal cities in the least developed parts of the world. However, it is precisely these areas that currently feature no planned eco-cities and little eco-urban planning aimed at adaptation to climate change and mitigation of its effects. Thirdly, the geography of eco-cities raises questions about the extent to which we are witnessing the development of these cities as yet another configuration of socio-economically uneven practices of gating (Newman, Beatley and Boyer's 'divided city' scenario on a vast scale), through the construction of new 'green' enclaves for those who can afford to live there, with little or no concern for wider society.

Holding the eco-city to account

It is in the landscape of a panoply of global, diffuse and socially constructed and determined crises that eco-cities are being proposed

DOI: 10.1057/9781137298768.0004

as bounded, neat solutions to all of these problems, depending on the context in which the blueprints for these cities are being elaborated. The current focus on proposing and developing eco-cities is a deeply modern endeavour, with roots in 19th and 20th century utopian urban planning. Eco-city projects are proposed as master plans for visions of social, environmental, technological and economic change which are largely based on technical knowledge and global knowledge elites (from the urban planning elites of firms such as Foster + Partners, to developers such as Arup, who display equal ability in engineering visions of the city as they do in imagineering and marketing these visions). The imaginaries of eco-urbanism are, in turn, rooted in deeply ecologically modernising policies and approaches to environmental-economic regulation that view the union of market, technology, and expert knowledge as a way of overcoming limits to growth.[40]

As such, eco-city projects are modern visions shot through with a strong tendency to enshrine specific 'ways of doing' in the new urban arena, and to commodify the environment. Most contemporary eco-city plans are for normative urban areas which materialise ways of doing and knowing: it is at this juncture that eco-cities can most appropriately be called utopian.

When it comes to commodifying the environment, the centrality of the link between technology and modern visions of utopia in the eco-city is not surprising: technology is a set of 'ways of doing' upon which the modern project is founded, and the eco-cities of the early 21st century are largely based on processes which in turn segment and commodify 'nature' and the environment into usable and marketable products for urban consumption.[41] These processes of commodification are most visibly focused, in the eco-city, on borderless resources such as the sun, air, or water.

Commodification of this kind is most visible in networked infrastructure, such as solar panel arrays, wind towers, and hydroelectric installations or solar-powered desalination plants.[42] A focus on technology within the context of modernity can thus enable a more grounded understanding not only of the broad societal landscape within which eco-cities are being planned, financed, and built. The eco-city *as an urban technology* in itself also becomes a useful way of conceptualising the ways in which nature and society are being (re)configured in new urban contexts justified through recourse to totemic (albeit rooted in undoubtedly sound technical knowledge) notions of crisis.

DOI: 10.1057/9781137298768.0004

The focus of this book is on a critical interrogation of eco-city projects: in order to do this, it focuses in detail on two distinct new-build projects. Through the examples of Tianjin eco-city, China, and Masdar eco-city, Abu Dhabi, the following analyses the two largest eco-cities under construction at the time of writing. In doing so, it attempts to move beyond the plans, blueprints, and architects' and planners' visions for these new cities of the future. The book considers the contexts in which these eco-cities were designed and planned: in both cases, within less than deliberative and open planning and govern-ance regimes; with a strong focus on city development and delivery by corporations; and with still-extant questions around how to build cities so as to ensure social as well as environmental and economic sustainability.

Studying new urban areas such as Tianjin and Masdar eco-cities opens up possibilities for discussion and debate around the increasing preva-lence of eco-cities as new, emergent forms of eco-urbanism. It is impor-tant to keep in mind that the eco-city reflects key concerns prevalent in the Age of Crisis, such as those identified at the start of this chapter. Likewise, the Garden Cities planned at the end of the 1800s can be seen as responses to the deep sense of unease that followed the development of the modern industrial city. And yet, while many of the Garden Cities and New Towns of the 19th and 20th centuries contained elements of planning for socially progressive (albeit sometimes questionable and, in hindsight, often fatally flawed) aims, the level to which contem-porary eco-cities include a concern with progressive socio-economic planning within their remits is rather less apparent. It is perhaps with misplaced nostalgia that urban scholars today consider cities such as the turn-of-the-century British paternalistic industrial towns of Bournville, or Port Sunlight, with their careful provision for workers and their fami-lies. Yet, in today's eco-urbanism, there is so little concern with workers and with those who do not belong to the socio-economic elite, that nostalgia is an understandable response.

Garden Cities and paternalistic industrial towns did not, of course, scratch very far below the socio-economic surface of modern urban-industrial capitalism to challenge the economic system which produced the sorts of unhealthy and negative urban industrial envi-ronments which prompted socially-minded thinkers and planners to envision Garden Cities and the like in the first place. It is also outside the scope of this book to delve in detail into the deep workings of

DOI: 10.1057/9781137298768.0004

contemporary capitalism. However, a discussion of eco-cities as they are being planned and built today can be a useful entry point into a wider discussion about the economic-industrial system which produced the environmental externalities which eco-cities are meant to 'solve'. This is a crucial discussion to be had, because otherwise urban scholarship risks focusing on symptoms rather than underlying conditions, on what is visible rather than on the systemic (mis)functioning of political-economic life which materialises deeply unjust but highly visible urban environments.

However, as shown in the second and third chapters, few eco-city projects seem to touch upon issues of socio-economic equity and justice in much detail. There is little sense, especially in some of the shinier corporate visions for grand eco-cities, of transition planning which moves past specific visions of the city as an environmental-economic sphere. Examples such as Tianjin and Masdar eco-city can be seen as valiant attempts to fashion new, 'green' urban areas in the context of rapidly changing environmental conditions, but the accounts presented in the second and third chapters stress the need to consider the eco-city's socio-economic aspects as well as transition planning which is integrated across different scales and which stretches beyond the necessarily narrow remit of individual eco-cities.

At the time of writing this book, contemporary eco-city developments, particularly of the new-build type, often seem to raise more questions than those they set out to answer in the first place. And yet, eco-city planning today can also be seen as a great opportunity, since the eco-urban trend within which eco-cities are situated is essentially a project aiming to throw the city wide open and to re-imagine what urban life in the near future could look like.

Today's eco-cities are attempts to bring about particular visions of the ideal city. Clearly, many of the visions that have thus far been capturing the attention of the world's media are capital-intensive, technocentric plans which utilise notions of crisis to justify the roll-out of what can be described as elite visions of the city. Therefore, to what extent are eco-cities simply *technical* experiments where the *social* is an afterthought? Nonetheless, it would be easy to limit analysis of eco-cities to mere critique. However, this would be a wasted opportunity, and the final chapter in this book aims to highlight some alternative ways of thinking about and building eco-cities which move past a fetishising focus on high-tech solutions in a social vacuum.

DOI: 10.1057/9781137298768.0004

Notes

1   Harvey, D. (1996) *Justice, Nature and the Geography of Difference*. Bognor Regis, Wiley.

2   Castells, M. (2010) *End of Millennium, 2nd Ed*. Oxford, Wiley-Blackwell. See also the other volumes in the *Information Age* trilogy by Manuel Castells. Reed, T. V. (2014). *Digitized Lives: Culture, Power and Social Change in the Internet Era*. London, Routledge.

3   Giddens, A. (1990) *The Consequences of Modernity*. Oxford, Polity Press-Balckwell: 2.

4   Fukuyama, F. (1992) *The End of History and the Last Man*. New York, NY: Free Press.

5   Maalouf, A. (2012) *Disordered World: A Vision for the Post-9/11 World*. London, Bloomsbury Paperbacks.

6   Dodds, K. and Ingram, A., eds. (2009) *Spaces of Security and Insecurity: Geographies of the War on Terror*. Farnham, Ashgate.

7   Beck, U. (1992) *Risk Society*. London, Sage.

8   Foster, J. B. (1997) The crisis of the Earth: Marx's theory of ecological sustainability as a Nature-imposed necessity for human production. *Organization & Environment* 10: 278–95.

9   Mol, A.P.J. and Spaargaren, G. (2009) Environment, modernity and the risk-society: the apocalyptic horizon of environmental reform. *International Sociology* 4: 431–59; Swyngedouw, E. (2010) Apocalypse forever? Post-political populism and the spectre of climate change. *Theory, Culture and Society* 27: 213–32.

10 Bulkeley, H. and Castán Broto, V. (2012) Government by experiment? Global cities and the governing of climate change. *Transactions of the Institute of British Geographers* 38: 361–75.

11 Newman, P., Beatley, T. and Boyer, H. (2009) *Resilient Cities: Responding to Peak Oil and Climate Change*. Washington, DC, The Island Press.

12 Newman *et al.*, *Resilient Cities*, 37–8.

13 Newman *et al.*, *Resilient Cities*, 44.

14 Newman *et al.*, *Resilient Cities*, 51.

15 Discussion of urban resilience and of the resilient city is necessarily limited in this introduction. Discussions of the idea of resilient cities with a focus on security can be found in Coaffee, J. (2009) *Terrorism, Risk and the Global City: Towards Urban Resilience*. Farnham, Ashgate. For urban resilience with regards to environmental change see Watson, D. and Adams, M. (2011) *Design for Flooding: Architecture, Landscape, and Urban Design for Resilience to Climate Change*. Hoboken, NJ: John Wiley & Sons; and Gasparini, P., Manfredi, G. and Asprone, D., eds. (2014) *Resilience and Sustainability in Relation to Natural Disasters: A Challenge for Future Cities*. New York, Springer;

DOI: 10.1057/9781137298768.0004

Leichenko, R. (2011) Climate change and urban resilience. *Current Opinion in Environmental Sustainability* 3(3): 164–8.

16 Suzuki, H., Dastur, A., Moffatt, S., Yabuki, N. and Maruyama, H. (2010) *Eco2 Cities: Ecological Cities as Economic Cities*. Washington, DC: The World Bank.

17 Ecocity Builders website (2014) Ecocity Builders. Available at: http://www. ecocitybuilders.org Accessed 1 March 2014. See also Joss, S., Tomozeiu, D. and Cowley, R. (2012) Eco-city indicators: governance challenges. In Pacetti, M., Passerini, G., Brebbia, C. A. and Latini, G. (eds.) *Sustainable City VII: Urban Regeneration and Sustainability*. Southampton, WIT Press: 109–20.

18 Isles, M. (2014) Melbourne's eco city transition plan. Local Government Managers Australia, National Office website. Available at: http://www.lgma. org.au/default/melbournes_eco_city_transition_plan Accessed 19 August 2014; Ministère du Logement et de l'Égalité des Territoires (2014) Les EcoCités. Available at: http://www.territoires.gouv.fr/les-ecocites Accessed 19 August 2014.

19 Ruano, M. (1999) *Eco-Urbanism: Sustainable Urban Settlements*. Barcelona, Gustavo Gili.

20 Bulkeley, H. (2013) *Cities and Climate Change*. London, Routledge. See also Hoffman, M. J. (2011) *Climate Governance at the Crossroads: Experimenting with a Global Response*. Oxford, Oxford University Press.

21 Geels, F. (2005) Co-evolution of technology and society: the transition in water supply and personal hygiene in the Netherlands (1850–1930) – a case study in multi-level perspective. *Technology in Society* 27: 363–97; Geels, F. W. and Verhees, B. (2011) Cultural legitimacy and framing struggles in innovation journeys: a cultural-performative perspective and a case study of Dutch nuclear energy (1945–86). *Technological Forecasting and Social Change* 78(6): 910–30.

22 The body of work which focuses on understanding the mechanisms through which transitions happen is known as the Multi-Level Perspective (MLP). See Geels, F. W. (2002) Technological transitions as evolutionary reconfiguration processes: a multi-level perspective and a case-study *Research Policy* 31(8–9): 1257–74.

23 Geels, F.W. and Schot, J.W. (2007) Typology of sociotechnical transition pathways. *Research Policy* 36(3): 399–417; Bailey, I. and Wilson, G. (2009) Theorising transitional pathways in response to climate change: technocentrism, ecocentrism, and the carbon economy. *Environment and Planning A* 41(10): 2324–41.

24 Coenen, L., Benneworth, P. and Truffer, B. (2012) Toward a spatial perspective on sustainability transitions. *Research Policy* 41(6): 968–79.

25 Brown, H. S. and Vergragt, P. J. (2008) Bounded socio-technical experiments as agents of systemic change: the case of a zero-energy residential building. *Technological Forecasting & Social Change* 75: 107–30.

DOI: 10.1057/9781137298768.0004

26 Geels, *Co-evolution of technology and society*.

27 Castán Broto, V., Glendinning, S., Dewberry, E., Walsh, C. and Powell, M. (2014) What can we learn about transitions for sustainability from infrastructure shocks? *Technological Forecasting and Social Change* 84: 186–96. For analysis of how a 'shock' such as the Fukushima tsunami and associated nuclear disaster impacted on Japanese energy policy, see Elliott, D. (2012) *Fukushima: Impacts and Implications*. Basingstoke, Palgrave Macmillan.

28 Chien, S. S. (2013) Chinese eco-cities: a perspective of land-speculation-oriented local entrepreneurialism. *China Information* 27: 173–96.

29 Bulkeley, *Cities and climate change*.

30 Register, R. (1987) *Ecocity Berkeley: Building Cities for a Healthy Future*. Berkeley, CA: North Atlantic Books. See also the excellent discussion of the development of the eco-city concept in Roseland, M. (1997) Dimensions of the eco-city. *Cities* 14: 197–202.

31 Roseland, *Dimensions of the eco-city*.

32 Engwicht, D. (1992) *Towards an Eco-City: Calming the Traffic*. Sydney, Envirobook; Engwicht, D. (1993) *Reclaiming Our Cities and Towns: Better Living With Less Traffic*. Gabriola Island, BC, Canada: New Society Publishers.

33 Roelofs, J. (2000) Eco-cities and red green politics. *Capitalism Nature Socialism* 11: 139–48.

34 Kargon, R. H. and Molella, A. P. (2008) *Invented Edens: Techno-Cities of the Twentieth Century*. Cambridge, MA: MIT Press.

35 Roseland, *Dimensions of the eco-city*.

36 Castán Broto, V. (2012) Social housing and low carbon transitions in Ljubljana, Slovenia. *Environmental Innovation and Societal Transitions* 2: 82–97.

37 Joss, S., Tomozeiu, D. and Cowley, R. (2011) *Eco-Cities: A Global Survey 2011*. London, University of Westminster.

38 Joss et al., *Eco-Cities*.

39 Joss et al., *Eco-Cities*.

40 Hajer, M. (1995) *The Politics of Environmental Discourse: Ecological Modernization and the Policy Process*. Oxford, Oxford University Press; Mol, A.P.J. and Spaargaren, G. (1993) Environment, modernity and the risk society: the apocalyptic horizon of environmental reform. *International Sociology* 8: 431–59.

41 Brey, P. (2003) Theorizing modernity and technology. In: Misa T. J., Brey, P. and Feenberg, A. (eds.) *Modernity and Technology*. Cambridge, MA, MIT Press: 33–71.

42 Monstadt, J. (2009) Conceptualizing the political ecology of urban infrastructures: insights from technology and urban studies. *Environment and Planning A* 41(8): 1924–42.

DOI: 10.1057/9781137298768.0004

2
Experimental
Eco-Cities in China

Abstract: *Eco-urbanism, the aim of shaping cities to be more environmentally sensitive, is becoming increasingly popular in the People's Republic of China. The chapter introduces Chinese eco-urbanism by contextualising it within the broader remit of concern over environmental, demographic and economic crisis which has seen Chinese planners and policymakers concerned with an array of issues such as rapid rural-urban migration, environmental despoliation and the question of how to transition towards a higher-value economy. These issues are largely urban in focus, and the rest of the chapter critically considers eco-urban responses to these crises in the largest new-build eco-city project in the world: the Sino-Singapore Tianjin Eco-City.*

Caprotti, Federico. *Eco-Cities and the Transition to Low Carbon Economies.* Basingstoke: Palgrave Macmillan, 2015. DOI: 10.1057/9781137298768.0005.

Introduction: eco-urbanism in China

Eco-urbanism is increasingly prevalent in the Chinese urban landscape. From low-carbon communities, to eco-industrial parks and individual green buildings, the country's planners and architects are gradually turning their attention to the question of combining sustainability with the country's current love affair with concrete, steel, and glass. Against this background, eco-cities are fast becoming a reality in China: urban scholars point to the hundreds of new 'eco' urban areas on the drawing board throughout the country.[1] Several projects are already under construction while some were, at the time of writing, nearing completion. Naturally, eco-city development on a scale as large as that found in China has generated project failures as well as successes. Some eco-cities have stalled and been put on hold, while some high-profile developments have fallen by the wayside and have been shelved: these include Dongtan, a planned large eco-city near Shanghai, and Huangbaiyu, a partially-completed model eco-village project in Liaoning province.[2]

China's concern with eco-city development cannot be considered in isolation from the emergence of eco-urbanism, in various guises, throughout East and South-East Asia. While some scholars of urban China seem to consider the contemporary Chinese city through a largely exceptionalist lens due to the country's current hyper-urbanisation, China's concern with greening urban development is part of a wider, global movement aiming to think of cities as experimental and as locations for techno-economic and environmental transitions, as seen in the previous chapter. The Asian context is, arguably, one of the most vibrant laboratories in which new, innovative city plans and urban technologies are being trialled, and new towns and eco-city developments are central to this trend.[3] Examples of the many projects underway or completed at the time of writing, include: Songdo, near Seoul in South Korea, a city combining elements of Smart City and eco-urban planning; Lavasa, a sustainable city in India; Cyberjaya, part of Malaysia's Multimedia Super Corridor; and the 26 cities categorised as 'eco-towns' in Japan since 1997.[4]

Within this context of continuing interest in 'green' cities, and more specifically within the setting of emergent eco-urbanism in Asia, China is rapidly becoming a leader in the number of eco-city projects being developed, in the broad scope of eco-city policy and planning, and in the range of technological solutions with which planners are experimenting.

DOI: 10.1057/9781137298768.0005

Apart from its current focus on eco-cities, China has a long tradition of experimental urbanism. In the 20th century, this has included experimentation with cities as building blocks of national industrial-economic plans and trajectories during the Mao era: during the period from 1949 to 1978, cities were reconfigured and subject to a degree of scrutiny, control, and national coordination never before seen in Chinese history. Cities became part of a highly problematic, countrywide, communist experiment with mobilising the forces of industrial production for accelerating national development. Some of the forms of urban organisation rolled out during the period to 1978 have had a lasting impact on the city, from the metamorphosis of pre-communist 'wall and gate' examples of urban form into 'work units' (*danwei*), to the situation of heavy industries close to centres of habitation (a strategy whose implications are felt to this day), to the establishment of the *hukou* system of residency registration which ties citizens to specific rural or urban areas.[5] In the post-Mao period, urban experimentation has continued apace. This has included the use of Special Economic Zones (SEZs), within which economic reforms could be trialled.

These zones have had knock-on consequences in existing cities throughout China (especially in coastal areas), but they have also been the places where new types of cities have emerged: Shenzhen, in the Pearl River Delta, is perhaps the most well-known example of this phenomenon, although by no means the only specimen of SEZ-driven urban growth.[6] Other experimental forms of urbanism have included the construction of college towns, as well as industrial parks, such as those built in Suzhou and Wuxi in the 1990s in collaboration with Singapore.[7] Many of these projects have been heavily criticised for their overarching focus on economic return and because of the issues associated with their potential for generating revenues for local governments. In the case of Jiangnan, a college town near Chongqing, for example, Chien Shiuh-Shen, a geographer at National Taiwan University, has highlighted how:

> [L]and designated for education was re-zoned as commercial land for housing and commerce. In this case, farmers were paid RMB 20,000 per *mu*.[8] Chongqing University then leased the land to Jiangnan, the land developer in charge of the Chongqing project, at a price of RMB 400,000 per *mu*. For that purchase, Chongqing University netted a startling RMB 380,000 per *mu*. In addition, Jiangnan used the land for commercial developments, housing, resorts, hotels, offices, and so on, thereby earning further profit.[9]

DOI: 10.1057/9781137298768.0005

Nonetheless, it is clear that the contemporary focus on eco-cities as experimental urban areas is, in a sense, nothing new: as seen above, Chinese policymakers, urban and economic planners, and architects have used the city as an experimental zone for over 60 years. Eco-cities are simply the latest attempt to engage with the urban arena in ways designed to deliver desired visions of economic as well as urban development, and of 'sustainable' urban areas and more desirable levels of quality of life. While there are clearly delineated geographies to these experimental projects (in particular, Chien notes that while college towns can mostly be found near China's coasts, eco-cities are more evenly spread out, including in China's western and central regions), the development of eco-cities can be seen as the latest in a long line of urban experiments in China.[10]

Crisis and the Chinese city

The eco-city projects built in China in the 2000s and 2010s can be considered the latest in a long genealogy of experimental urban schemes. However, one of the characteristics that set them apart from previous urban experimental developments is not their focus on technology, or on transitional economic strategies, or even on 'green' architecture and design. Rather, what distinguishes the current multiplication of eco-city projects throughout the mainland is the fact that eco-cities in contemporary China have been designed as a response to, and within a wider context of, crisis.

The Age of Crisis is expressed in specific and detailed ways in different geographical locations and national contexts: this is to be expected, as different localities, cities and states will concentrate on the specific crises which are identified as most directly affecting these places and their associated socio-political groups. In the Chinese context, the distinct set of crises that has been used to justify the rollout of eco-city projects across the country can be segmented into environmental, demographic, economic, and social crises. With regards to environmental crisis, the increasing despoliation and contamination of China's terrestrial, aerial, and aquatic environments as a result of the country's breakneck industrial and economic development are well known. Environmentally, it is clear that China's rapid industrial revolution has had significant and multiple impacts. These include the production of negative externalities

DOI: 10.1057/9781137298768.0005

as a result of industrialisation, and as a consequence of the rapid urbanisation of formerly rural areas. An example of this is the phenomenon of 'cancer villages,' hamlets and towns where rates of specific cancers are very high and which happen to be located close to the heavily polluting, often poorly regulated and carcinogenic industries which are characteristic of the heart of China's industrial engine.[11]

Another highly visible example of the urban environmental impacts of industrialisation is the decrease in air quality in cities throughout the country. In early 2013, record-high levels of smog in cities such as Beijing led commentators to talk about an 'airpocalypse', or 'airmageddon' in Chinese cities.[12] This has sparked widespread concern over the extent to which the negative environmental impacts of economic development are actively harming the country's citizens, especially its urban dwellers. This is due in large part to the invisible but potentially life-shortening risks of breathing high concentrations of small particulate matter (also known as PM2.5, particles which are up to 2.5 micrometres in size), which can settle in the lungs and enter the bloodstream.[13]

Apocalyptic visions of Chinese cities choked by their own industrial and automotive emissions abound, especially at times of record levels of smog, highlighting the perception of environmental risk as diffuse, invisible, and yet ever present. This is exemplified by the proliferation and popularity of air quality indicators, both online and in the press. It is also typified by the growth of a market for goods aimed at protecting individuals from urban smog: products range from apartment air purifiers, to smog masks, to apps which enable the user to know the real-time particulate levels in their city. For example, a China Air Quality Index app is sold through Apple's iTunes store, and is available for download for both iPad and iPhone; and websites such as the Air Quality Index China site (http:aqicn.org) provide city-level, real-time information of the concentrations of pollutants from PM2.5, to ozone and others. In addition, the website sells pollution masks and hosts pages urging China's Ministry of Environmental Protection (MEP) for stricter controls on coal-based pollutants. From industrial emissions to the individualisation of environmental risk in the growing Chinese city, concern over environmental crisis has been a key characteristic of an increasing amount of public and policy awareness and apprehension over the pace and scale of the environmental despoliation which is part of the long shadow projected by China's economic miracle.

DOI: 10.1057/9781137298768.0005

In demographic terms, the now-slowing but still rapid rate of rural-urban migration is seen in some quarters as a continuing crisis which impacts cities' growth and development plans. While migrant labour fuels the growth of China's cities, migrant workers' integration into urban economies and societies has been a key concern for decades. In addition, the demographic shifts which have seen a rebalancing of population in China away from rural areas and towards cities have highlighted the environmental and socio-economic injustices and inequalities which have accompanied China's urban growth. These inequalities range from oppressive work practices, to barriers to accessing services that only urban dwellers (those who hold the *hukou* residency permit for a specific city) are entitled to, to obstacles for the children of migrants in terms of schooling, education, health, and other elements of social life.[14]

Nonetheless, it is widely acknowledged that the range of issues folded into the idea of demographic crisis is wide and encompasses issues other than rural-urban migration. As Feng Wang, director of the Brookings-Tsinghua Centre argues, the population crisis in China relates to three distinct issues: the increasing numbers of migrants moving to the cities (160 million according to Wang), the country's rapidly ageing profile, and the gender imbalance among the young as a result of the approximately 13 million abortions (many of them gender-specific) which take place *every year* in the country as a result of its One Child Policy.[15]

All of these factors – migration, ageing, and an increasingly imbalanced and pathologised reproductive landscape – have become a serious cause for concern with regards to the shape of China's future cities. Indeed, the demographic crisis is *already* having an effect on cities, and on the country as a whole. Just consider the fact that, due to low birth rates and the abhorrent use of abortion as a demographic policy instrument, the 750,000 primary schools that existed in China in 1990 decreased in number to around 300,000 in 2008. In terms of ageing, low rates of population growth mean that China is seeing the most rapid rise in the proportion of the population aged over 60 ever recorded in world history. As Wang points out,

> Such a compressed process of demographic transition means that, compared with other countries in the world, China will have far less time to prepare its social and economic infrastructure to deal with the effects of a rapidly ageing population.[16]

DOI: 10.1057/9781137298768.0005

The overall urban picture gets even more complicated when considering the effect of changing socio-economic and cultural norms in conjunction with the environmental effects of demographic crisis. As geographer Jared Diamond and sustainability scholar Jianguo Liu argued in the *Nature* journal in 2005, even though China's population growth rate has rapidly slowed, in 1985–2000 the number of households in the country increased at a rate which was three times faster than the population growth rate. This was because household sizes decreased over the same time period, from an average of four-and-a-half people per household in 1985, to three-and-a-half in 2000. Liu and Diamond also pointed to the increasing numbers of divorces (1.6 million in 2004 alone) in the country, which are exacerbating housing and environmental pressures. This is because divorces effectively double the number of housing units required, reduce average household size, and have environmental effects due to the fact that two households consume more energy, on average, than a single household housing the same number of people.[17]

When considering China's economy, the concept of 'crisis' is not often mentioned. After all, China's meteoric rise is generally described in superlative terms. Its constantly high year-on-year GDP growth figures have been widely admired: since the market-focused reforms which began in 1978, the country has performed in a stellar fashion, rising from the world's tenth largest economy in 1982, to the second largest from 2002 onwards, according to the World Bank's historical GDP figures. Nonetheless, a key concern is the extent to which the economy can successfully transition away from its current engine of growth and towards new, high value-added forms of economic growth with a focus on environmental and other technologies, services, and products.

This, coupled with several other existing economic problems, paints a variegated picture of economic performance, even as the wider macro landscape is one of still-stellar GDP growth (albeit now slower than the double-digit growth that characterised China's economy in the first decade of the 21st century). The other economic issues facing the country include concerns over internal debt, and anxieties about overheating urban property markets, leading to questions not about a single housing 'bubble,' but about a series of bubbles in different cities and municipalities. The urban consequences of the rise of real estate bubbles are clearly serious, as exemplified by the collapse of the housing market in Wenzhou, Zhejiang province, in 2010. Speculative practices and the flipping of residential units led to a market that was so overheated that residential

DOI: 10.1057/9781137298768.0005

property was, at the height of the bubble, priced at up to US $11,000 per square metre. This figure was worth around double the average wage in Wenzhou at the time. In 2010, the central government's imposition of curbs on property purchases and more stringent requirements on mortgage down payments caused a collapse in Wenzhou residential prices. In a localised version of the West's mortgage crisis, the average price of properties fell by about 60 per cent between 2010 and 2012.[18]

One of the key crises facing China's policymakers and urban planners today lies at the juncture of environmental, demographic, and economic issues: the *social* crisis facing China's urban arena. Worsening urban environmental conditions, a changing and imbalanced demographic context, and rising income and socio-economic inequalities for both migrants and city dwellers form a potentially unstable mix, of which China's political leaders are well aware.

Indeed, it is in the context of a rapidly shifting socio-economic landscape in the country's cities that the government introduced its policy aim of building a 'harmonious society' as part of China's Eleventh Five-Year Plan (2006–10). This broad aim, introduced and backed by then-President Hu Jintao, is an attempt to form an amalgam between social sustainability and stability and a continued focus on technological and economic development. The harmonious society is meant to be the backbone of a stable polity. The challenges facing the construction of a truly harmonious society are daunting: the realms of finance, education, health care, and social welfare are riven with inequalities in terms of access, levels of provision, and entitlements.[19] When many of the country's approximately 160 million migrant workers have difficult or no access to health care, and when health care is priced above what is affordable for low-wage individuals, questions of social sustainability abound.[20] Furthermore, a key question is whether local governments effectively interpret calls for a 'harmonious society' as promoting measures to stifle dissent and silence critical voices: as Kin-man Chan has argued, the harmonious society concept 'can be interpreted as praise for absolute social order and used as a convenient tool for suppressing dissent'.[21]

The environmental, economic, demographic, and social problems highlighted above constitute a geographically specific configuration of the broader, global concerns that characterise the Age of Crisis. These problems, risks, and hazards, from toxic emissions to ensuring stable urban societies, are issues that have not just concerned scholars and the

DOI: 10.1057/9781137298768.0005

public; they are keenly felt at the highest level of politics and policymaking.[22] For example, in the case of environmental degradation, Pan Yue, a vice minister at China's Ministry of Environmental Protection, sounded a warning call almost a decade ago, in 2005.[23] He argued that China's economic miracle 'will end soon because the environment can no longer keep pace'.[24]

Entrepreneurial eco-cities

It is against the background of a set of interlinked crises that the eco-city model can be most adequately situated in the Chinese context. Indeed, while eco-cities are, on the surface, directly focused on achieving more ecologically and environmentally amenable urban development trajectories, their promise goes deeper than this. Eco-city projects are being proposed as new urban areas that can be designed and planned so as to provide successful solutions to wider questions around economic and social development and the achievement of a harmonious society.

Eco-cities are one of the experimental means through which the future shape of the Chinese city is being trialled and tested. They are by no means the first attempt to engage with delivering greener urban environments: China's late 1990s emphasis on green community (*luse shequ*) initiatives with a focus on individual behaviour at the residential level is a case in point.[25] Rather, what is new about the multiple eco-city projects being built in the country today is, firstly, their holistically experimental nature: eco-cities in today's China are not merely attempts to construct a single edifice or a set of buildings, but wholesale attempts to re-imagine the whole city, its economy, and its society. This places them apart from other forms of eco-urbanism, such as green building, which focus on very specific and circumscribed aspects of the urban environment, and mostly from an architectural, environmental and economic point of view. Eco-cities across China are being used as bounded experimental containers within which the wide range of approaches, technologies, and planning approaches found within the broader eco-urban trend can be applied and tested on a large scale.

Secondly, Chinese eco-cities are characterised by their *transitional* nature. In this sense, they are conceived as potential niches for the testing and eventual introduction of market reforms aimed at kick-starting transitions towards a low-carbon, green economy. However, they are

DOI: 10.1057/9781137298768.0005

also transitional in a social sense, due to the fact that they are often planned with the objective of constructing societies that are harmonious – although what that will mean in practice is open to debate.

Thirdly, Chinese eco-cities are distinctive because they can be seen as *entrepreneurial* cities. This is because, by and large, government authorities and the private sector are both involved in defining eco-city visions and goals. Today's eco-cities are seen as opportunities for investment as opposed to simply centrally-planned, government-ordained visions of a greener and more socially stable urban future. At the same time, as Fulong Wu, a professor of planning at University College London, has argued, eco-cities can be seen as the direct result of the entrepreneurial nature of municipal and other forms of local government in China. This is because most eco-city projects are funded not directly from government funds but by real estate developments: as Wu argues, '[eco-city] development is mainly a corporate operation, although these projects may be supported by the local state through facilitating and speeding up the process of going through project approval and other regulatory controls. China's buoyant housing market makes these projects financially viable because, through investing in a development project, house price inflation may create a profit when the properties are sold.'[26]

The involvement of state and private actors in designing, developing, delivering, and profiting from these projects thus enables eco-cities to be seen as entrepreneurial cities. In order to understand eco-cities as examples of urban entrepreneurialism, it is key to place the concept of the entrepreneurial city and of entrepreneurial city governments within an economic-historical context rooted in the specific configurations of the land tenure system and property market in China since Mao's death. This is because it is key to understand the state-land nexus in China in order to understand the phenomenon of eco-city construction.

The spatial consequences of Mao's death included a gradual relaxing of state control over land, to the extent that the Constitution was amended in the 1990s to allow local authorities to lease land.[27] This meant that local governments started to re-designate land from rural to urban use, so as to be able to lease it and thus finance infrastructure and other projects with the proceeds. A consequence of this was that between 1999 and the late 2000s, land-related revenue at the local level rose from 10 to 55 per cent of total local budgets. By the end of the 2000s, land conveyance fees were responsible for covering 80–100 per cent of the total cost of urban infrastructure projects.[28]

DOI: 10.1057/9781137298768.0005

In 2005, the central government instituted a farmland preservation policy that meant that local governments were only able to convert a limited amount of rural land to urban classification. The policy (the '1.8 billion *mu* farmland threshold') was aimed at preserving rural land while also curbing the excesses of local government in converting land for financial purposes. This is one of the reasons for the failure of the Dongtan eco-city project, scheduled for construction on Chongming Island, near Shanghai. The Shanghai city government could not reclassify Chongming Island as urban because doing so would have breached the Shanghai area's conversion quota. Chongming Island had been designated as protected farmland previously, in part to enable Shanghai municipality to remain within its conversion limit. The inability to reclassify the island from rural to urban was a key driver of the project's failure. Conversely, the farmland preservation policy can be seen as the main driver for the selection of a marshland as the site for the Sino-Singapore Tianjin Eco-City. The site was a non-arable salt marsh, unsuitable for designation as rural land. It was, therefore, easy to reclassify for urban usage. The marshland site is conveniently located near Tianjin, where the city government was facing similar land conversion limit pressures to those faced by Shanghai.[29]

The enthusiasm for eco-city and other urban and infrastructure projects in China cannot be adequately understood in isolation from the broad national trend towards land-based entrepreneurialism and speculation, which is in turn based on the recent reconfiguration of the state-land nexus. In addition, new urban projects are skilfully marketed by local governments and interests so as to reflect national priorities. For example, 1990s college town developments were depicted as responses to national initiatives to form a knowledge-based economy. In the late 2000s, eco-cities became depicted as a way of responding to national priorities around ecological modernisation and a greener, more sustainable circular economy.[30]

Most studies of entrepreneurial cities in the Chinese context focus squarely on the link between land, the state, and the private sector in designing and delivering urban and infrastructure projects, from industrial parks to eco-cities. This is an important emphasis, as the revenue potentially generated from urban development is one of the key drivers of the hyper-urbanisation of contemporary China. However, one of the aspects of entrepreneurial eco-city building that is often overlooked in accounts of urban entrepreneurialism is the fact that several eco-city projects are being built, at least partly, as instruments of what could be loosely termed foreign policy. This is apparent when viewing the international partnerships which characterise many flagship eco-city developments, from the joint venture with the Singapore government to develop Tianjin eco-city, to the involvement of a range of government and other actors from the Netherlands in the Sino-Dutch Shenzhen Low Carbon City project, to cite just two examples.

Singapore is, at the time of writing, the state that has established the longest track record in developing Sino-foreign urban projects. In part, this is due to the fact that Singapore is, in the words of economist Meine Pieter van Dijk, 'a kind of laboratory for housing and environmental policies in Asia'.[31] Apart from its involvement in Tianjin eco-city, Singapore has been centrally involved in the older Suzhou Industrial Park development, as well as the Guangzhou Knowledge City project. Both Tianjin eco-city and Suzhou Industrial Park saw the participation of both countries' central governments, while Guangzhou Knowledge City is the result of a partnership between Singapore and the Guangdong provincial government.[32] The significant investment (in political and financial terms) by Singapore in projects in China, such as Suzhou Industrial Park, is an indicator of the utility of these projects in obtaining and securing political goodwill from the Chinese government. For example, a steering committee meeting was held in October 2013 between the Chinese and Singaporean partners responsible for developing Tianjin eco-city. This saw the involvement of Singapore's deputy prime minister Teo Chee Hean, and of Zhang Gaoli, China's vice-premier. The meeting was held on the same day as the tenth Joint Council for Bilateral Cooperation between the two countries, at which Singaporean firms were authorised to invest more heavily in Chinese securities. The Council meeting also saw discussion on improving 'measures to promote cross-border flows of Yuan between China-Singapore Suzhou Industrial Park as well as the Tianjin Eco-City'.[33] Clearly, meetings to discuss eco-city developments as well as bilateral relations were closely intertwined, and Singapore's Chinese industrial and urban projects were central to policy discussions.

Therefore, it is clear that eco-city developments in China can be interpreted through an entrepreneurial lens, not only in terms of land tenure and conveyance for local governments, or as profitable opportunities for development corporations, but also in some cases as strategic initiatives in the context of international and government-to-government relations.

DOI: 10.1057/9781137298768.0005

The next section focuses in more detail on the international Sino-foreign eco-city project at the most advanced stage of construction at the time of writing: the Sino-Singapore Tianjin Eco-City.

Sino-Singapore Tianjin Eco-City: planning for transition

Rising from the site of a former salt marsh near Tianjin port, the Sino-Singapore Tianjin Eco-City is one of the largest construction sites in China. Cranes, wind turbines, and rapidly emerging forests of residential buildings populate the skyline, while six-lane highways circumscribe as-yet empty blocks of wasteland designated for future residential development. Tianjin eco-city is one of China's flagship urban projects. As a central government-backed project, it is a city that cannot (and will not be allowed to) fail. As such, it is both an exception to the urban norm, populated as it is by cities subject to the vagaries of market and socio-economic forces, and a fascinating window into a particular vision of China's urban, ecological, and social future. The fact that Tianjin eco-city carries the stamp of the Chinese government means that the project effectively serves as an indicator for future urban development in the country. The visions of urban life that are being developed for, and in, the new city are visions of what an ecologically and socially harmonious city will look like over the next decade or so.[34] As to how the city will function in actuality, and whether the construction of physical urban environments can successfully go hand-in-hand with the development of socially sustainable and just communities and neighbourhoods, that remains an important open question and one that will be discussed at the end of this chapter.

The international scale: China, Singapore, and the global market

Tianjin eco-city is an international joint venture, as highlighted in the previous section. This means that the broadest scale at which the eco-city project can be considered is the international one. Indeed, the new city was international in character since its inception: the project was proposed following a call by the central government in April 2007 for a landmark eco-city to be built. While the process was somewhat competitive – Tianjin, Tangshan, Urumqi, and Baotou municipalities

DOI: 10.1057/9781137298768.0005

all submitted bids – Tianjin seemed to be a clear favourite due to its potential high-level international links. When Tianjin was announced in November 2007 as the winner of the site selection process for China's flagship eco-city, Singapore's involvement as a joint venture partner was crucial. Tianjin's bid built on, and extended, existing links between the two countries in the construction of urban and industrial projects. What was also key was the location of the eco-city within the Tianjin Binhai New Area (TBNA), a Special Economic Zone (SEZ). The TBNA is administered by the Tianjin Economic-Technological Development Area (TEDA), the first state-sponsored technological development zone in northern China.[35]

The TBNA was instituted in 2006 as part of the central government's strategy to promote economic development in pilot zones where industrial and commercial reforms and initiatives could be tested.[36] A series of economic inducements, including corporate income tax incentives for high-tech firms, were deployed to stimulate industrial and commercial development in the area. The economy within the SEZ developed rapidly, reaching a record 24 per cent annual growth rate in 2009. This growth rate is reflected in the economic performance of Tianjin's metro area, where the economy grew at an average of 16 per cent in 1999–2009.[37] Both the TBNA and TEDA are significant interfaces between Chinese industry and manufacturing, and the international market. As such, the location of the Sino-Singapore Tianjin Eco-City within the TBNA is a significant pointer as to the outward-facing role of the project.

Tianjin eco-city's international nature is also evident in its organisational structure. The institution through which the joint venture between China and Singapore is operationalised is the Sino-Singapore Tianjin Eco-City Investment and Development Corporation (SSTECIDC). The corporation is jointly owned by a Chinese consortium led by the Tianjin TEDA Investment Holding Company, and a Singaporean consortium headed by the Keppel Corporation, a real estate development firm. The organisational and planning system through which the city is being developed is highly hierarchical, reflecting the tiers of stakeholders active in the project. At the highest level, a yearly meeting takes place between the premiers of China and Singapore. Other meetings, involving government ministers working on national construction projects, occur once or twice per annum. At the municipal level, meetings take place several times a year, between local city officials from Tianjin municipality and executives from private corporations such as development firms. At this

DOI: 10.1057/9781137298768.0005

level, the meetings' specific outputs include reports and plans for specific buildings within the eco-city.

The regional scale: the Bohai Rim Megalopolis

On a regional scale, Tianjin eco-city can be considered within the broad context of the coastal area of north-east China's Bohai Rim. The Bohai Rim is made up of the provinces of Hebei, Liaoning, and Shandong, as well as Beijing and Tianjin municipalities. Demographically, the region contains around 18 per cent of China's population, which is reflected in its heavily urbanised character. The Bohai Sea's coasts feature some of China's largest urban agglomerations, including Tianjin, Tangshan, Shenyang, and Qingdao, as well as others. It is this that has prompted some scholars to call the Bohai Rim, with its now interlinked cities, the 'Bohai Megalopolis'.[38]

Economically, the Bohai Rim is the third most important region in China after the Pearl and Yangtze River deltas. As a result of policies promoting heavy industries in the area during the Mao era, coupled with pro-market reform policies after 1978, the region is heavily industrialised. Environmentally, the Bohai Rim's economic and demographic importance has led to significant environmental and ecosystem deterioration: as a team of environmental scientists at Tsinghua University, Beijing, has argued, 'natural resource shortages and pollution have been caused by the incompatibility of heavy and chemical industry aggregation with a sustainable environment'.[39] This meant that by 2007 the Bohai Rim had exceeded its estimated environmental carrying capacity by circa 36 per cent.[40] This state of affairs is also reflected in the urban environment. In 2013, two of China's most polluted cities were part of the Bohai Megalopolis. Beijing and Tianjin were, respectively, the second and sixth most polluted cities in the country.[41]

The Bohai Rim Megalopolis region is thus of fundamental importance to the central government's focus on facing up to the crises identified above: the region sits at the confluence of economic-industrial, demographic, economic, and environmental trends which are increasingly casting cities such as Tianjin as the stages on which crisis – or its solutions – will play out.

The interest in using new-build eco-cities as transitional niches in the context of the Bohai Rim Megalopolis can be seen in the existence of more than one large-scale international eco-city project in the region.

DOI: 10.1057/9781137298768.0005

Apart from Tianjin eco-city, an ambitious international eco-city was planned for further north along the coast, near the city of Tangshan's port zone.[42] The eco-city, named Caofeidian, is on hold at the time of writing, but it is a sign of the significant regional interest in green urbanism and transitional policies (and potential real estate-related profit). The successful development of transitional initiatives and projects such as Tianjin eco-city can therefore be seen in the light of a wider, highly urbanised region in which the eco-city can potentially play the pivotal role of a socio-technical niche around which successful transitional trajectories can form.

The city scale: Tianjin municipality

Tianjin is China's third-largest city, after Beijing and Shanghai: it is one of the country's most politically important cities, especially in administrative terms. Tianjin municipality has a similar status and level of importance as a province, being one of China's five 'National Central Cities,' together with Beijing, Shanghai, Chongqing, and Guangzhou. The central importance of Tianjin as an urban area can also be seen in the fact that it is one of just four cities in the whole country that are under the direct control of the central government.[43]

Demographically, the city's population of nearly 13 million has grown rapidly, especially since 2000. The city's population in 1982 was around 7.7 million. In the 1980s, Tianjin added a further million residents to its population; in the 1990s, it added slightly more than a million. In the period from 2000 to 2010, as the city's economic trajectory started to skyrocket, the population grew at three times the pace of the previous two decades, adding around three million people over the course of the decade. It is telling that by the end of 2010, one in six inhabitants of Tianjin (totalling between one and a half and two million people) did not hold a Tianjin *hukou*.[44]

Thus, while the city is situated within a wider regional context of ongoing industrialisation, population pressures, and environmental concerns, it also faces challenges specific to the urban area in and around Tianjin. The selection of an area of wetland close to the Bohai Sea near Tianjin for a new eco-city mega-project can be seen as a grand trial in designing a city as a bounded experiment in which 'solutions' to all of the above challenges can be tested and verified.

DOI: 10.1057/9781137298768.0005

Planning and eco-urban design: land, capital, and nature

In order to explore the use of the Sino-Singapore Tianjin Eco-City project for developing and proposing solutions to 'crisis,' as well as promoting transition towards a low-carbon economy, it is useful to start from some of the basic components that are central to the eco-city. These are the *land* on which the city is being built and on whose value the entrepreneurial nature of the project is predicated; the involvement of private and other forms of *capital* in financing, driving and enabling the project; and the type of *nature* which the eco-city claims to be protecting and fostering.

Land

The master plan for the Sino-Singapore Tianjin Eco-City is an attractive and multi-coloured, zoned diagram showing the eco-city's main residential and commercial areas. The project site is a large area, around 30 square kilometres in extent. Most of the land on which the eco-city is being developed was a wetland that included some salt farming areas, although the city will also feature a lake. This is a reclaimed body of water, formerly used as a pond for effluent discharge. When reclamation is complete, the pond will be known as Qingjing Lake. The city itself is bisected by the meandering Old Ji Canal, a 1,000-year old waterway which flows out of the course of the new Ji Canal at the northern end of the eco-city, meanders through the site, and then re-joins the canal's main course at a more southerly location on the project site. Overall, the land chosen for the eco-city is not zoned as agricultural or urban land, which made it relatively straightforward to appropriate the large site for reclassification for urban use.

The benefits of identifying 'unproductive,' non-agricultural lands for urban development were mentioned above. In the case of Tianjin eco-city, the potential financial benefit (for Tianjin municipality and real estate development firms) of the transformation of a former wetland into a high-end urban area is clear. For real estate developers, the ability of purchasing plots of land at relatively cheap prices and then selling apartments at a profit is, understandably, a key driver.

For example, in 2013 the Keppel Corporation purchased a 10.37 Ha parcel of land near the Ji Canal for RMB 241.1 million (US $49.1 million). This means that the plot was valued at around RMB 2,300 per square

DOI: 10.1057/9781137298768.0005

metre. The corporation was reported to be planning to build 350 apartments in high-rise residential blocks for 'upper-middle income homebuyers'.[45] Since the average price of real estate in the eco-city is around RMB 9,000 (US $1,449) per square metre, and since each square metre of land is worth several times its asking price due to the fact that apartments occupy the same vertical column on a specific land area, the returns on investment can be sizeable. The development is expected to comprise 350 homes, as well as retail premises and three office towers. Positing a conservative estimate of average apartment size at 135 square metres (for a high-end apartment) means that the potential average price of such an apartment could be around RMB 1.21 million (US $195,595).[46] If all 350 apartments were to sell at this price, the real estate corporation would earn around RMB 425.25 million (US $68.46 million), before the leases and sale values for the three office towers and residential premises are calculated. Thus, the financial driver for developing plots of land in the eco-city is clear.

Therefore, one of the key *material* determinants of the development of Tianjin eco-city is not crisis per se, but China's land market, as well as the country's development regulations. This means that land deemed unproductive or non-agricultural has a market value close to zero; the potential for profit, if this land is developed, is therefore very high. In this context, the injection of land into the eco-city development framework is the first step in the development of entrepreneurial eco-city developments. This is a process which effectively commodifies land, and which privatises and markets it. Thus, 'valueless' land is turned into an investment multiplier by reclassifying it and transferring it to real estate corporations for development. The fact that all land remains the property of the Chinese state means, in conceptual terms, that China's government ultimately owns the site. However, the complex web of real estate development licenses, land leases, and agreements between states and corporations, as well as between China and Singapore themselves, mean that land in an experimental eco-urban project such as Tianjin eco-city has been effectively turned into a commodity.

This opens up questions around the sustainability of these practices. In the Chinese context, the link between eco-cities and land speculation has become so glaring as to prompt planner and scholar Elizabeth Rapoport to declare, in a recent review of eco-city projects, that in China, 'most eco-cities are first and foremost entrepreneurial land development projects... While they may be sincerely committed to sustainability, economic growth and cost minimisation are the prevailing concerns of Chinese authorities.'[47]

While there are specifically Chinese characteristics to this trend, recent developments farther afield have raised similar questions, as with the UK government's announcement of a small Garden City project in Ebbsfleet, in the Thames Estuary near London, which has raised concerns as to the construction of high-end eco-housing as part of the project, as well as to the use of a former quarry site, in a flood risk area, on which to build the eco-city. Nonetheless, land and the way it is valued in the Chinese market and by the Chinese state is clearly a key determinant of the attractiveness of Tianjin eco-city for both government authorities and municipalities, and for development corporations and investors.

Capital

In the case of Tianjin eco-city, it is clear that the city is a deeply entrepreneurial project. The state was involved in a primary driving role, initiating the project and facilitating site selection, as well as enabling the treatment of the eco-city as an experimental zone in terms of policies aimed at a transitional economy. However, as seen above, the project's entrepreneurial nature can also be seen in the involvement of international corporations that aim to profit from involvement in the project. This is the case at the level of the consortia responsible for the joint venture through which the project is structured, as well as at the level of the involvement of private companies in delivering the city's infrastructure, residential quarters, and associated physical urban environments and technologies.

With regards to the former, the Chinese consortium not only involves a 20 per cent stake by the China Development Bank, one of China's three policy banks directly controlled by government ministers, it also involves a 45 per cent stake financed by private investors. The Singaporean consortium involves a 50 per cent stake held by the Keppel Corporation, as well as a 10 per cent investment by the Qatar Investment Authority and a 40 per cent stake held by several private investors. Thus, even at the level of the governing consortia, the Tianjin eco-city project is characterised by the involvement of several corporate actors as well as state investment vehicles.

In terms of the involvement of private actors in building the eco-city, the overall picture is one of directed entrepreneurialism. The city itself is divided into large residential 'blocks,' which lend themselves well to the segmentation of the eco-city plan into areas to be developed by

DOI: 10.1057/9781137298768.0005

different private corporations. This has resulted in a range of different residential block projects by development companies. As well as the Keppel Corporation, the eco-city's residential and commercial areas are being developed by corporations such as Taiwan's Farglory; Shimao of Hong Kong; Mitsui Fudosan, from Japan; Sunway, from Malaysia; and Vanke and Vantone, both from China. Thus, a key part of the eco-city's entrepreneurial nature is the existence of a range of links and partnerships between private corporations and government actors.

Nature

The Tianjin eco-city project is clearly rooted in the specific ways in which land is valued, and in which value can be extracted from commodifying, privatising, and marketing the land on which the city is being built. At the same time, it is predicated on domestic and cross-border flows of capital, enabled by bilateral government relations and entrepreneurial municipal authorities. These flows of capital (in the forms of investment as well as human, social, cultural, and symbolic forms of capital) in turn enable the materialisation of the eco-city and the transformation of the wetland into an urban area.

The engineering of wetland and salt marsh into a steel and glass eco-city has, at its root, the modern ideal of the science-based transformation of nature. Tianjin eco-city can be seen as yet another example in a long line of socio-natural engineering projects which stretches from at least the 19th century to the present time, and which are socially constructed as 'solutions' to a range of 'crises'. From the Paris sewers bringing hygiene and morality to the modern metropolis, to embankments disciplining rivers and nature in capital cities such as London, to the heavy infrastructure of water networks in Europe, North America, and Asia, to the myriad new cities built in the 20th century, nature has been contested, reshaped, and transformed into something more 'positive', in line with the discourse of modern 'civilisation', and economic-technological progress.[48] The emphasis on rationality, technology, and progress as the means through which nature could be destroyed, changed, and replaced with a more desirable environment is a key characteristic of modernity.[49]

On the Tianjin eco-city site, salt marshes and wetland environments, laced with toxic contaminants from China's drive to industrialise, represent a specific, bounded experimental location where the transformation of local, environmentally despoiled conditions into a utopian,

DOI: 10.1057/9781137298768.0005

green environment can be successfully achieved. This highlights, again, the importance of the eco-city as an *experimental* site, both materially and symbolically. Materially, the eco-city is important as a specific site where change can be effected through capital, technology, and ecologically modernising approaches to the transformation and urbanisation of the wetland environment. Symbolically, Tianjin eco-city represents an identifiable signifier of success in facing and 'solving' the crises which are seen as facing urban China. Thus, the success of the Tianjin eco-city project can be understood as symbolising the success of the deployment of elite engineering and technological knowledge, coupled with entrepreneurial governments and the market, in delivering environmentally, socially, and economically amenable results.

One of the ways in which the deeply modern destruction of a 'negative' nature existing before the eco-city and its transformation into a 'positive' eco-urban environment can be seen in the representation of the wetland on which the eco-city is being built. The eco-city project is described as being founded on land defined as unusable, with urban development taking place in the face of a hostile nature. The land is appropriated and 'converted' to 'positive' use, whereby 'positive' signifies a specific use value by selected users; potential target residents, businesses, real estate corporations, investors, and the governments of China and Singapore, among others. A host of technologies, from land reclamation to green building, become the means through which the initial negative nature of a partially contaminated salt marsh is turned into a green and (in marketing documents, at least) luxuriant, harmonious eco-city.

The resemblance between the portrayal and celebration of this high-tech, 21st century eco-city project and major hydro-engineering projects in the early 20th century (from land reclamation in the Netherlands, to the draining of Italy's Pontine Marshes and the construction of New Towns on the reclaimed land) is striking. Consider the way in which the *New York Times* commended the reclamation of the Pontine Marshes by Italy's fascist regime in a 1932 article:

> Six years ago the Fascist Government undertook to transform the pestilential Pontine plain into Italy's model agricultural region. Undaunted ..., a handful of devoted engineers and an army of sturdy workers, equipped with the latest machinery, attacked the stagnant waters of the plain, determined to triumph with the aid of modern engineering ... They were heard from only intermittently until a few months ago, when it was announced that the reclaiming of a portion of the Pontine Marshes was an accomplished fact... A tangible sign of

DOI: 10.1057/9781137298768.0005

the redemption of the Pontine Marshes is the inauguration of the new city of Littoria, the future capital of the region.[50]

The role of technology and engineering in destroying a negative 'first nature' and replacing it with a positive, modern 'second nature' is clearly celebrated: indeed, the above article highlights the first New Town built in the reclaimed marshes as a symbol of modernity's overcoming of natural limits and constraints.[51] Compare the description quoted above, with the same newspaper's celebration, 79 years later, of the reclaiming of the land on which Tianjin eco-city is being built:

> Three years ago, this coastal area fit perfectly into the dictionary definition for 'wasteland.' Its soil was too salty to grow crops. It was polluted enough to scare away potential residents. Sometimes the few fishermen who lived here saw investors driving in, but they quickly turned around and left, leaving nothing behind except dust.
>
> But then some people showed up to buy a piece of this land. It is about half the size of Manhattan. They restored the soil, cleaned up water pollution and began preparing the once-deserted place for a city that will host green businesses and some 350,000 residents by 2020.
>
> This is Tianjin Eco-City, basically a wasteland-to-community experiment carried out by the Chinese and Singapore governments.[52]

The site the of effluent discharge pond which is being turned into Qingjing Lake is a prime example of the type of remediation work which is taking place on the eco-city site, and which sees technology as the central pivot of transformation of a negative environment into a more 'positive' one. The pond was built in the 1970s and collected residential and industrial wastewater which was then released into the Ji Canal to subsequently flow into the Bohai Sea. Both the water and the sediments at the bottom of the pond were highly contaminated by a range of pollutants, including DDT and heavy metals such as mercury, arsenic, copper, and cadmium. Plans to turn the contaminated pond into the central landscape feature of Qingjing Lake included dredging the pond, filling tubes made of high-strength, permeable textiles with the dredged slurry, and then allowing the water from the slurry to drain out from the tubes. The tubes were then laid in layers next to the lake, capped with a geomembrane and coarse-grained fill, and topped off with a thick layer of topsoil. The mound of contaminated sediments was turned into a 12 hectare, nine metre-high mound which functions as a 'green' landscape feature.[53] Royal Ten Cate, the corporation responsible for the textile

DOI: 10.1057/9781137298768.0005

tubes into which slurry was placed and dewatered, described the process as enabling the pond to turn into a 'pristine lake'.[54] The mound will
clearly require continued evaluation and monitoring to prevent leaching
of contaminants and harmful substances. Thus, through the application
of technology, the eco-city site can be seen as effectively internalising the
waste from the area's previous use.

Planning for eco-urbanism

The Sino-Singapore Tianjin Eco-City is a project built at the confluence
of factors involving land, capital, and nature. It is also a city built as an
experimental way of tackling the Age of Crisis as it is perceived and
expressed in China today. In part, this is done through the top-down
planning framework which is being used as a guide for development of
the new city.

The master plan for Tianjin eco-city is one of the ways in which the
city's planning and design has been visualised by city planners and engineers. However, one of the other key mechanisms through which the
city is being used as an experiment to build solutions to the various environmental, demographic, economic, and other crises identified above is
through a set of 'scientific,' technical indicators which were devised as
drivers for the development of the project.

The indicator system proposed for Tianjin eco-city is, at first sight,
impressive. It is composed of a set of Key Performance Indicators (KPIs),
which provide targets for the eco-city, and benchmarks against which
the project's performance can be assessed. Several eco-cities worldwide
plan to use indicator and evaluation frameworks tailored to the specific
city in question, and according to the preferences of those who develop a
particular eco-city project. In the case of Tianjin eco-city, the Singapore
and Chinese governments elaborated 22 quantitative and four qualitative
KPIs.[55] Many of these are highly ambitious, such as the KPI that states that
by 2020, all buildings in the city should be classed as 'green buildings,' or
the one that states that by 2020, 90 per cent of journeys in the city should
be 'green trips'. There is also an impressive-sounding goal of having
20 per cent of the energy used in the eco-city derived from renewable
sources, again by 2020. Some of the quantitative indicators relate to the
city's knowledge/technical base: one KPI states that the eco-city should
feature at least 50 engineers or researchers per 10,000 workers.

DOI: 10.1057/9781137298768.0005

It is understandable and commendable that the eco-city should have sustainability targets. It is also important for projects such as Tianjin eco-city to be assessed during the project construction and development phase: benchmarks and measurable indicators are a key element of the eco-city experiment. However, there are several critiques that can be made of indicators as they are embedded within projects such as Tianjin eco-city.

Firstly, a key question is how indicators are developed, by whom, and with what level of participation by key actors and stakeholders. It is evident that Tianjin eco-city's indicators were developed as scientific and technical standards and measurements by scientists, engineers, and city planners, thus forming a set of technocentric indicators over which any potential citizens or any wider community have little say. Thus, indicator frameworks risk becoming tools of technocracy, defined as the rule of technical and bureaucratic elites over wider society.

Secondly, it is important to question to what extent indicators present 'empty' promises of sustainability. For example, the KPI that requires Tianjin eco-city to derive 20 per cent of its energy from renewables by 2020 is notable. However, China has a national target of achieving 15 per cent of national energy production from 'renewable' sources, which the government defines as including large hydropower projects and nuclear power.[56] By the end of 2009, however, officials expected the country to be able to achieve 20 per cent of its energy output from 'renewable' sources by 2020.[57] Thus, if national targets were to be achieved, Tianjin eco-city could do little or nothing to power itself with renewable energy, and still achieve its 20 per cent renewable energy KPI by simply plugging into the national grid.

Thirdly, it is key to ask questions as to how indicators of *social* sustainability are being included in the project's assessment and evaluation framework. As stated above, the eco-city is meant to provide a model 'harmonious' urban society once the project is completed. However, the eco-city's KPIs are mostly focused on technical, environmental standards. The overarching focus on quantification and environmental KPIs misses the important point that making a city environmentally sustainable and highly technological does not ensure utopia: it potentially risks the opposite.

With regards to the eco-city's built environment, the buildings being constructed on the site of the new city all achieve minimum standards of green building. The standard used for green building in Tianjin eco-city

DOI: 10.1057/9781137298768.0005

was developed by hybridising China's Green Star system with Singapore's Green Mark standard. The resulting Green Building Evaluation Standard (GBES) is being used to drive the requirements for residential and commercial buildings in the eco-city: it is based on a three-star system of rating the 'greenness' of buildings. In addition, in order to stimulate the green building market throughout China, the country's Ministry of Housing and Urban-Rural Development (MOHURD) announced in 2011 that it would subsidise the construction of three star-certified green buildings at a rate of RMB 75 (US $12) per square metre, thus providing an incentive to real estate and commercial property developers to enter the highest-rated green building market. This is a positive development, since by 2011 there were only 200 buildings in the whole of China which were certified as three-star, and most of these were government buildings.[58] Furthermore, underlining the entrepreneurial nature of the eco-city, it is interesting to note that global professional services firm PricewaterhouseCoopers (PwC) estimated, in a report on the Chinese green building market, that gaining a green building certification can mean a return of up to 9.9 per cent more per building for a real estate developer.[59]

The adoption of a green building standard will help the city achieve respectable standards of energy usage and energy intensity, and will ensure that residential as well as commercial units will be better insulated and feature more environmentally-friendly technologies than their counterparts in contemporary Chinese cities. In addition, the GBES used in the Tianjin eco-city project is different, because it incorporates more project-focused features, than the national green building standard, which is also (confusingly) known by the same acronym. In particular, the Tianjin GBES includes elements that the national standard system does not incorporate. Many of these elements show that the development of the GBES was sensitive to the local environment, such as the length of daylight in the Tianjin area during different seasons, and desired temperature differences between indoor and outdoor environments.

Green building and infrastructures have been made highly visible by the inclusion of micro-renewable technologies throughout the city. These range from solar water heaters displayed outside the eco-apartments in the city, to micro-solar panels and micro-wind turbines used to power street lighting and other visible technological elements in the city. The city also features less visible 'green' infrastructures, such as a centralised water purification system which means that drinking water from the tap should be safer than in contemporary Chinese cities. This fact is a key selling point in eco-apartment marketing, a fact readily ascertained when visiting apartments in the eco-city, since one of the first things real estate agents do with prospective property buyers is to offer them a glass of water from the tap. This points to one of the project's key characteristics: Tianjin eco-city is designed as a 'green' and 'sustainable' urban area, but one of its main selling points is that 'eco' means 'safe' urban living for its inhabitants as well. For example, the marketing for Farglory's 'U-City N. 1' residential development carries the slogan *Life @ U-City: Pretty Healthy & Cozy*, highlighting the need for the new city to be a healthy environment for urban dwellers worried about pollution and invisible contaminants, and concerned about investing not just in property, but in domestic spaces which can ensure (through clean drinking water, and even purified air through clean air filters) a healthy living environment. Alana Boland has called this phenomenon the emergence of 'filtered communities' in China: the provision of 'clean' urban environments through privatised infrastructures and technologies (such as purified water systems), in response to diffuse concerns and fear about the personal impacts of environmental crisis:

> Fear played a much more central role in the creation of these water systems such that in buildings and housing estates with piped purified water, protection from that which is 'bad' became a type of bounded public good. By adding a new hydrological layer to the gated form, premium water supply helped to create filtered communities. It represented a new mode of provision as well as protection – a mode that was underwritten by a consumerist approach to the distribution of environmental protection and public goods.[60]

It could be argued that the whole of Tianjin eco-city, through its provision of a separate water system, as well as technologies such as domestic air filters, and its broader targeting of wealthy residents, can be considered a 'filtered' or 'purified' community.[61] Or, more precisely, every building and residential block, and indeed every eco-apartment, can be seen as bounded, constituent parts of the filtered community which is the experimental Tianjin eco-city. This, in turn, opens up the question of whether the eco-city will continue to exist as a filtered, premium eco-enclave, or whether it will be able to become a transitional niche which will enable the construction of greener and more sustainable urban environments in a wider, regional and domestic context.

Useful indicators of this will be, for example, whether the more stringent green building standards adopted in the eco-city project can feed

DOI: 10.1057/9781137298768.0005

into national green building standards and certification processes, or whether the project's KPIs can influence other cities' urban development pathways.

It will take decades to assess the impact of a project such as Tianjin eco-city on the national urban landscape. Alternatively, there is a risk that the eco-city will remain a green oasis in the midst of the rapidly deteriorating environment of the Bohai Megalopolis. It is an inescapable irony that contemporary eco-cities often appear as 'green islands' within wider, environmentally dystopian landscapes. Some scholars have queried whether eco-cities are just attractive, new urban areas in the midst of industrial 'junkscapes': Tai-Chee Wong, at Nanyang Technological University, Singapore, has called them 'pearls in the sea of degrading urban environments', benefiting those who can afford to live there.[62] Hopefully, Tianjin eco-city will not fall into this category in environmental terms. Whether it does so in terms of its social makeup is the focus of the rest of the chapter.

The question of public housing and white-collar visions

One of the eco-city's aims is for the project to include at least 20 per cent of its residential units as public housing. This is an admirable aim, especially in a context where private developers are clearly driven by the aim of maximising profits on housing units. In this sense, developing regulations for the inclusion of public housing in the eco-city works towards the 'harmonious society' ideal, as well as ensuring urban social sustainability through enabling socio-economic diversity.

However, there are key questions to be asked about the extent to which public housing in Tianjin eco-city is *truly* public, in the sense of enabling low to lower-middle income households to locate in the eco-city, thus promoting true socio-economic mixing. The regulations for apartments designated as public housing state that only married couples who don't already own a residential unit in the eco-city can apply for public housing, in recognition of the need to promote population of the city by households, not lone individuals, and not by those who may profit by property speculation on other apartments in the city. Furthermore, public housing units in Tianjin eco-city are sold furnished, which reduces moving-in costs for residents.[63]

DOI: 10.1057/9781137298768.0005

Some of the other regulations for public housing make the notion of 'public' seem more restrictive. Firstly, access to public housing in the eco-city is restricted to *either* those who have worked in the eco-city for more than three years, *or* those who hold a non-agricultural *hukou* and have worked in the eco-city for more than a year. This last requirement effectively excludes Tianjin's nearly two million migrants who hail from rural areas (and who therefore do not hold an urban *hukou*) from owning public housing units in the eco-city. Secondly, pricing for public housing is fixed at around RMB 7,000 (US $1,127) per square metre compared to an average of RMB 9,000 (US $1,449) per square metre in the eco-city's other housing units.[64] Although both prices are much cheaper than the average RMB 20,000 (US $3,220) per square metre in Tianjin's downtown area, the price differential partially reflects the fact that the eco-city is in fact around 40 kilometres from the downtown area, and is much closer to Binhai. Thirdly, the public housing regulations for the eco-city stipulate that those applying for this type of housing should not be earning more than RMB 60,000 (US $9,659) per year. However, this maximum earning figure is around 30 per cent higher than the average yearly wage in Tianjin municipality, meaning that those earning higher than average wages will be able to apply for public housing in the eco-city. As Cecilia Springer, a US Fulbright Program researcher formerly based at Tianjin's Nankai University has argued:

> In general, Eco-City housing caters to a more affluent clientele than average – white-collar workers from the Binhai New Area, the TEDA Start-up Area, and leaders and officials involved in the Eco-City project. While the public housing claims to resolve the housing problems of middle to low income families, it also is aimed at a reasonably wealthy demographic.[65]

The question of how the eco-city will develop socially and in terms of the formation of urban communities remains open to question. What is evident is that the project's focus is, to date, largely technocratic and reliant on technology and design as ways of achieving the deliverables requested by the KPI framework that is being used to develop and assess the eco-city. In terms of social sustainability, for example, the eco-city relies on the framework of the United States Green Building Council (USGBC) Leadership in Energy and Environmental Design for Neighborhood Development (LEED-ND) standard, which is largely focused on urban design, land use and environmental standards as ways

DOI: 10.1057/9781137298768.0005

of shaping a sustainable city both environmentally and socially. As sociologist Yifei Li has argued:

> Technocracy is most evident in the case of Tianjin. The use of KPIs throughout the planning and construction stages effectively 'blackboxed' much of the discussion of the project details. Its subscription to LEED-ND standard also makes discussion of community development fairly technical. In other words, the evidence suggests that the [...] state has exhibited tendencies towards technocracy [...].[66]

Such concerns seem very distant from the images of eco-city life offered in marketing and advertising materials by SSTEC and real estate development corporations. Understandably, such visions focus less on KPIs and more on the potentially positive urban environment being built in the eco-city. When browsing the marketing material proffered to potential customers by real estate agents in and around Tianjin eco-city, it becomes clear that the city is being marketed to a specific clientele. The visions of eco-living that are presented in apartment brochures and on real estate websites are of aspirational, white-collar lifestyles in comfortable and relaxed surroundings. This opens up questions as to the extent to which the city is to be truly socially sustainable in terms of enabling a thriving and vibrant social life: or the extent to which the project will become an 'eco-enclave' for the wealthy.[67] This was already a concern with other prominent eco-urban projects. For example, the failed Dongtan eco-city near Shanghai was criticised because of its targeting of residents from Shanghai, while 'affordable' housing in the eco-city was deemed to be too expensive for local farmers to purchase.[68] Therefore, one of the questions that needs to be asked is the extent to which eco-cities such as the one near Tianjin are conceived (and perceived) not as ecologically sustainable and harmonious urban areas, but as desirable real estate destinations for aspirational urban dwellers.[69]

An example of the marketing of the eco-city to potential residents as a comfortable, wealthy urban environment is the way in which Vantone's six-story apartment developments in Tianjin eco-city are described as 'Western-style apartment houses'. These are residences that vary from 100 to 187 square metres in size, as compared to the neighbouring 17–18 floor residential apartment blocks, which feature slightly smaller residences (100–48 square metres). An advertising poster for this development, which is called 'Eco-City Legacy Homes,' shows a man dressed in business attire, relaxing on a sofa under the following slogan: *Comfort is*

DOI: 10.1057/9781137298768.0005

the ultimate aim of living. This highlights the fact that life in the eco-city is being conceived as a comfortable existence for white-collar workers.

Likewise, a saccharine advertising video produced by SSTEC, titled *Tianjin Eco-City, My Home,* takes the viewer through a hypothetical day in the lives of a household of eco-city residents. The couple, who live in a new apartment in the city (together with a son and grandparents), are shown enjoying distinctly elite lifestyles: the husband works in an office complex within the eco-city and walks to work, while his wife is described as a heavy user of Tianjin international airport for business trips to Japan and further afield. The whole video features sunny skies and lush, green natural surroundings: a far cry from the ever-present air pollution and lack of vegetation that are the reality in the area at the time of writing. Nonetheless, marketing and advertising materials such as these highlight the fact that the eco-city's private developments are being marketed as appropriate residential destinations for aspirational and elite workers.

One of the other aspects of the Tianjin eco-city project which points to the eco-city being a reworking (and perhaps an intensification) of existing socio-economic inequalities in urban China is the juxtaposition between the educational and other inequalities suffered by the millions of rural-urban migrants in Tianjin, and the availability of world-class educational facilities in the eco-city – at a price. While migrants struggle to gain equitable access to state education (for themselves or their offspring) as a result of the *hukou* curse which leaves them jurisdictionally tied to their rural town of origin, the new residents of the eco-city will be able to access excellent (and expensive) international schools.

The eco-city's inhabitants also live under the *hukou* system, but are presumably able to pay for private education and other facilities due to their high socio-economic status. For example, one of the kindergartens and elementary school facilities which have been set up in Tianjin eco-city is being co-developed by GEMS International, a United Arab Emirates-based education provider, with SSTECIDC. The resulting GEMS World Academy, which cost around US $18 million to finance, costs RMB 97,020 (US $15,618) *per year* to attend. However, the average yearly wage in Tianjin municipality was RMB 42,240 (US $6,800) in 2011, and the average wage of a migrant worker in China in 2013, according to China's Ministry of Human Resources and Social Security, was RMB 31,308 (US $5,098) per annum.[70] The fact that a private school charging more than double the average local citizen's yearly wages is choosing to locate in the eco-city is indicative of the eco-city project becoming identified as a zone for wealthy residents. Wealth thus enables the future constituency of the eco-city to rise above jurisdictional constraints and to access privatised public service delivery in the entrepreneurial eco-city.

One of the key ways in which Tianjin eco-city is seemingly developing into an 'eco-enclave' is discernible in the built environment, through the deployment of gating and walling throughout the city's residential areas. The residential blocks (or 'superblocks') which form the eco-city are large, measuring over 400 metres in length and 300 metres in width in some cases: more than double the average size of city blocks in Manhattan. These blocks feature high-rise residential towers interspersed with low-rise apartment residences, and are inspired by Singapore's residential areas, although the 'superblock' is a common element of urban morphology in most Chinese cities.[71]

Tianjin eco-city is meant to be a walkable city, and there are large pedestrianised areas within individual residential developments. However, several of the blocks also feature underground parking for residents' cars: this is often at the ground floor level, meaning that the whole residential block is effectively raised by a storey. The car parks have thin, barred windows to the outside, to enable air circulation. Walking along the sidewalk by the side of each block, it becomes apparent that the underground car parks' walls form an effective wall around the whole development, and on top of which the residential part of the block is built. This makes it impossible for a pedestrian to stray into the block except through the gated driveways that allow cars to access the block and its parking garages. It also means that the overall 'feel' of the eco-city's residential areas is of large, gated and sometimes walled blocks which a pedestrian can only access at certain pre-determined points, where the eyes of guards and other forms of surveillance can monitor movement.

One of the eco-city's KPIs states that 90 per cent of trips in the city should eventually be 'green'. The city's indicator framework defines 'green trips' as those which are non-motorised: walking, cycling, and public transport are included in the 'green trip' category. The eco-city will be served by public transport including a bus network and a light rail system, and electric taxis were also mooted for the city. However, presumably the KPI can be achieved by counting small journeys on foot between buildings in a single 'superblock' as a single 'trip' which does not cause any emissions.

DOI: 10.1057/9781137298768.0005

Notwithstanding the eco-city's aims of walkability and 'green' mobility, when examining the overall design of the city, it is evident that the project is being centred around the automobile. This is because while each residential 'superblock' is pedestrianised *within* the block itself, each of the blocks is separated from the next one by often very wide avenues. Some of the blocks in the city's residential Start-Up Area, for example, are separated by six-lane roads, with few crossing points for pedestrians. This, coupled with the difficulty of cutting across an urban superblock on foot due to walling and gating, makes potential trips on foot lengthy: the overall design of the city, with its multi-lane highways crisscrossing the residential parts of town, seems to resemble Le Corbusier's car-focused Plan Voisin for Paris rather than a human-sized urban plan sensitive to the need for city blocks to be permeable via multiple access points.

Thus, while Tianjin eco-city's experimental nature regarding KPIs, renewable energy, and the construction of a new urban environment to provide solutions to the concerns facing a rapidly urbanising China is replete with innovative ideas and technologies, the overall picture leaves many questions to be answered. Many of these questions are about aspects of the eco-city which indicator frameworks, technical analyses, and urban plans do not focus on in detail: namely, the extent to which the city will not only live up to its green targets and to its own marketing, but also the degree to which it will be enabled to become a more socially sustainable society. The marketing and construction of residential developments in Tianjin eco-city therefore leaves open the uncomfortable question as to whether the eco-city's projected future 'harmonious society' is not actually a project to construct what has been called a 'premium ecological enclave'.[72]

Planning for social sustainability and transition

The case of Tianjin eco-city shows some of the complexities and multi-faceted problems facing planners and policymakers as they confront the problems and crises thrown up by China's headlong drive to urbanise and develop its economy. Tianjin eco-city is currently a useful, if not altogether convincing, experiment in fashioning new urban environments and economies to be greener, more sustainable, and more supportive towards alternative visions of urban life.

DOI: 10.1057/9781137298768.0005

However, the discussion above has also highlighted how eco-cities such as the one in Tianjin must be analysed not just on their own terms, and by reference to their KPI frameworks and stated environmental achievements. Rather, a critical interrogation of eco-city frameworks must take place first and foremost in the context of the predominant ways in which land and nature are valued in a wider, national context: this provides useful clues as to what kind of nature is to be protected and valued in the eco-city, and what kind(s) of nature and environment are to be eliminated from new urban areas. In turn, a wider contextualisation also allows for an interrogation of the agency of governments and private capital in shaping visions of the city.

Scholars of urban China have been sounding cautions about the extent to which large-scale projects such as Tianjin eco-city are appropriate pathways for the development of socio-ecologically just urban environments in contemporary China. In her recent book *Urban China* (tellingly dedicated to China's migrant workers), Xuefei Ren of Michigan State University puts forward a strong indictment of the urban development pathway of which Tianjin eco-city is a flagship example:

> India, and probably other developing countries as well, have learned many wrong lessons from China, such as setting up Special Economic Zones, advocating massive investments in infrastructure, hosting mega-events, and pushing urban renewal by displacing the poor.[73]

Thus, strong and less than transparent state involvement in the delivery of urban projects, coupled with market and economic incentives for developers and other corporations, is increasingly seen by China scholars as essentially hollowing out the social and the political from new Chinese cities. This then raises the question of whether new eco-urban (and other) projects can really be said to be addressing the problems and crises they are designed to solve. To quote Ren again,

> [F]ollowing the Chinese model cannot solve these problems. As the Chinese experience shows, massive investment in infrastructure has put local governments in deep debt, and the shining new infrastructure projects often become profit-making machines for private-public partnerships. Hosting mega-events such as the Beijing Olympics has not brought many benefits to the people living in post-event cities, and the Shanghai style of urban renewal has displaced millions of the poor and turned inner-city neighborhoods into exclusive colonies for transnational elites.[74]

DOI: 10.1057/9781137298768.0005

In environmental terms, of course, the continued reclassification of rural land into urban expansion areas and new cities is of pressing concern. This is especially the case when, as in the case of Tianjin, wetland areas and land that is classified as non-productive are seen as blank slates on which construction can proceed. As environmental scholar and analyst Vaclav Smil presciently wrote in the early 1990s, there is often 'a clear invitation to set a plot of land aside as unsuitable for farming and then to build on it... [I]n countless cases all...regulations are simply ignored, and the transgressions are even turned into huge profits.'[75]

The construction of Tianjin eco-city thus leaves wide open the question of how to plan for *social* sustainability and a just socio-economic character in newly-planned and built urban environments. It is astonishing, but perhaps not surprising, that the planning of a whole new urban area such as Tianjin eco-city did not involve any of Tianjin's residents. This is not surprising in China, or anywhere else: from Ebenezer Howard's Garden Cities, to the UK government's stalled Eco-Towns initiative of the late 2000s, new town plans and ideas have by and large been characterised by elite planning and normative visions for what new urban environments should look like, what activities and ways of life they should prioritise, and how they should look. However, in an era in which deliberative and participatory planning is becoming ever more widespread, it would be desirable for eco-city plans to include an element of citizen participation in planning new urban environments. Furthermore, it is desirable for cities such as Tianjin eco-city to at least not perpetuate or even heighten existing socio-economic inequalities and disparities: it is far from certain that the eco-city will not exacerbate existing social problems, perhaps not within the city's confines, but certainly in terms of the difference in lifestyle and wealth between those living in the eco-city, those living outside it, and those who built it.

Notes

1 Wu, F. (2013) China's eco-cities. *Geoforum* 43(2): 169–71.
2 Baeumler, A., Chen M., Iuchi, K. and Suzuki, H. (2012) 'Eco-cities and low carbon cities: the China context and global perspectives.' In Baeumler, A., Ijjasz-Vasquez, E. and Mehndiratta, S. (eds) *Sustainable Low-Carbon City Development in China*. Washington, DC: The World Bank: 33–62.

DOI: 10.1057/9781137298768.0005

3   Keeton, R. (2011). *Rising in the East: Contemporary New Towns in Asia.* Amsterdam, SUN Architecture.

4   Kim, C. (2010) Place promotion and symbolic characterization of New Songdo City, South Korea. *Cities* 27(1): 13–19; Shwayri, S. (2013) A model Korean ubiquitous eco-city? The politics of making Songdo. *Journal of Urban Technology* 20(1): 39–55; Datta, A. (2012) India's ecocity? Environment, urbanisation, and mobility in the making of Lavasa. *Environment and Planning C: Government and Policy*, 30(6): 982–96; Brooker, D. (2012) "Build it and they will come?" A critical examination of utopian planning practices and their socio-spatial impacts in Malaysia's "intelligent city." *Asian Geographer* 29(1): 39–56; Van Berkel, R., Fujita, T., Hashimoto, S. and Geng, Y. (2009) Industrial and urban symbiosis in Japan: analysis of the Eco-Town program 1997–2006. *Journal of Environmental Management* 90(3): 1544–56.

5   Bray, D. (2006) Building 'community': new strategies of urban governance in China. *Economy and Society* 35(4): 530–49; Huang, Y. and Low, S. M. (2008) 'Is gating always exclusionary? A comparative analysis of gated communities in American and Chinese cities.' In Logan, J. R. (ed.) *Urban China in Transition*. Oxford, Blackwell: 182–202.

6   Yeung, Y-M., Lee, J. and Kee, G. (2009) China's Special Economic Zones at 30. *Eurasian Geography and Economics* 50(2): 222–40.

7   Bräutigam, D. and Tang, X. (2011) African Shenzhen: China's Special Economic Zones in Africa. *Journal of Modern African Studies* 49(1): 27–54.

8   A *mu* is a unit of area measurement equivalent to about 666.7 square metres, or 0.16 acres.

9   Chien, *Chinese eco-cities*, 184.

10 See Chien, *Chinese eco-cities*, 171, figure 1 for a map showing the more diffuse geographical distribution of contemporary eco-city project locations in China, compared with new college town locations.

11 Lora-Wainwright, A. (2010) An anthropology of 'cancer villages': villagers' perspectives and the politics of responsibility. *Journal of Contemporary China* 19(63): 79–99. See also Lora-Wainwright, A. (2013) *Fighting for Breath: Living Morally and Dying of Cancer in a Chinese Village*. Honolulu, University of Hawai'i Press.

12 Romano, G. (2013) No administrative solution in sight for urban 'Airpocalypse'. *China Perspectives* 3: 82–4.

13 Kaiman, J. (2013) Chinese struggle through 'airpocalypse' smog. *The Guardian*, 16 February 2013. Available at: http://www.theguardian.com/world/2013/feb/16/chinese-struggle-through-airpocalypse-smog Accessed 1 March 2014.

14 Gransow, B. and Daming, Z., (eds) (2010) *Migrants and Health in Urban China*. Berlin, Lit Verlag.

DOI: 10.1057/9781137298768.0005

15   Wang, F. (2010) China's population destiny: the looming crisis. The
       Brookings Institution, September 2010. Available at: http://www.
       brookings.edu/research/articles/2010/09/china-population-wang Accessed
       1 March 2014. See also Shan, J. and Qian, Y. (2009) Abortion statistics
       cause for concern. *China Daily*, 30 July 2009. Available at: http://www.
       chinadaily.com.cn/china/2009-07/30/content_8489656.htm Accessed 1
       March 2014.

16   Wang, *China's population destiny*.

17   Liu, J. and Diamond, J. (2005) China's environment in a globalizing world.
       Nature 435(30): 1179–86.

18   Leung, P. (2012) Wenzhou's once-hot housing market comes crashing down.
       South China Morning Post, 23 December 2012. Available at: http://www.scmp.
       com/news/china/article/1111003/wenzhous-once-hot-housing-market-has-
       come-crashing-down Accessed 1 March 2014.

19   Zhang, P. and Xu, M. (2011) The view from the county: China's regional
       inequalities of socio-economic development. *Annals of Economics and Finance*
       12(1): 183–98.

20   Milcent, C. (2010) Healthcare for migrants in urban China: a new frontier.
       China Perspectives 2010(4): 33–46.

21   Chan, K-m. (2009) Harmonious society. In Anheier, H.K., Toepler, S. and
       List, R. A. (eds) *International Encyclopedia of Civil Society*. New York, Springer:
       821–5, 824.

22   Economy, E. C. (2007) The Great Leap Backward? The costs of China's
       environmental crisis. *Foreign Affairs* 86(5): 38–59, 38.

23   The Ministry of Environmental Protection was formerly known as the State
       Environmental Protection Administration.

24   Yue, in Economy, *The Greap Leap Backward*, 38.

25   Boland, A. and Zhu, J. (2012) Public participation in China's green
       communities: mobilizing memories and structuring incentives. *Geoforum* 43:
       147–57.

26   Wu, *China's eco-cities*, 169.

27   Chien, *Chinese eco-cities*.

28   Chien, *Chinese eco-cities*.

29   Chien, *Chinese eco-cities*.

30   Chien, *Chinese eco-cities*.

31   van Dijk, M. P. (2011) Three ecological cities, examples of different
       approaches in Asia and Europe. In Wong, T.-C. and Yuen, B. (eds) *Eco-City
       Planning: Policies, Practice and Design*. New York, Springer: 31–50, 46.

32   de Jong, M., Yu, C., Chen, X., Wang, D. and Weijnen, M. (2013) Developing
       robust organizational frameworks for Sino-foreign eco-cities: comparing
       Sino-Dutch Shenzhen Low Carbon City with other initiatives. *Journal of
       Cleaner Production* 57: 209–20.

DOI: 10.1057/9781137298768.0005

33   Keppel Land (2013) *Across Borders.* 4th quarter, 2013. Keppel Corporation, Singapore, 12.

34   Caprotti, F. (2014) Critical research on eco-cities? A walk through the Sino-Singapore Tianjin Eco-City. *Cities: The International Journal of Urban Policy and Planning* 36: 10–17.

35   TEDA (2010) Tianjin Binhai New Area. Available at: http://en.investteda.org/BinhaiNewArea/default.htm Last accessed 1 March 2014.

36   Ministry of Finance and the State Administration of Taxation (2006) Incentives of the corporate income tax to support the development of the TBNA. 130(2006), 15 November 2006. Available at: http://en.investteda.org/download/revised20070115.doc Accessed 1 March 2014.

37   SSTECIDC. 2010. Celebrating eco. Available at: http://events.cleantech.com/tianjin/sites/default/files/SSTECBrochureFinal.pdf Accessed 1 March 2014.

38   Zhou, S., Dai, J. and Bu, J. (2013) City size distributions in China 1949 to 2010 and the impacts of government policies. *Cities* http://dx.doi.org/10.1016/j.cities.2013.04.011

39   Lin, L., Liu, Y., Chen, J., Zhang, T. and Zeng, S. (2011) Comparative analysis of environmental carrying capacity of the Bohai Sea Rim area in China. *Journal of Environmental Monitoring* 13(11): 3178–84, 3178.

40   Lin *et al., Comparative analysis.*

41   Na, L. (2013) Top 10 most polluted Chinese cities in 2012. *China.org.cn*, 15 April 2013. Available at: http://www.china.org.cn/top10/2013-04/15/content_28541619.htm Accessed 1 March 2014.

42   Joss, S. and Molella, A. (2013) The eco-city as urban technology: perspectives on Tangshan Caofeidian International Eco-City (China). *Journal of Urban Technology* 20(1): 115–37.

43   The other three municipalities under the direct control of the central government are Beijing, Shanghai and Chongqing.

44   Li, B. (2008) 'Why do migrant workers not participate in urban social security schemes? The case of the construction and service sectors in Tianjin.' In Nilsen, I. and Smyth, R. (eds) *Migration and Social Protection in China.* London, World Scientific: 184–204.

45   Keppel Land, *Across borders,* 13.

46   The apartment size estimate is based on the Keppel Corporation's existing Seasons Park residential development, where the high-end apartments are about 135 square metres in size.

47   Rapoport, *Utopian visions,* 142.

48   Gandy, M. (1999) The Paris sewers and the rationalization of urban space. *Transactions of the Institute of British Geographers* 24(1): 23–44; Kaika, M. (2006) Dams as symbols of modernisation: the urbanisation of nature between materiality and geographical representation. *Annals of the Association of American Geographers* 96(2): 276–301; Oliver, S. (2000) The Thames

DOI: 10.1057/9781137298768.0005

Embankment and the disciplining of nature in modernity. *The Geographical Journal* 166(3): 227–38; Swyngedouw, E. (1999). Modernity and hybridity: nature, *Regeracionismo* and the production of the Spanish waterscape, 1890–930. *Annals of the Association of American Geographers* 89(3): 443–65.

49 Almond, G. A., Chodorow, M. and Pearce, R. H. (1982) Historical, ideological, and evolutionary aspects. In Almond, G. A., Chodorow, M. and Pearce, R. H. (eds) *Progress and its Discontents*. Berkeley, University of California Press: 17–20.

50 Cortesi, A. (1932) The famous Pontine Marshes turned into smiling fields. *The New York Times*, 30 October 1932, XX4.

51 Cronon, W. (1991) *Nature's Metropolis: Chicago and the Great West*. London, Norton.

52 Liu, C. (2011) China's city of the future rises on a wasteland. *The New York Times*, 28 September 2011. Available at: http://www.nytimes.com/ cwire/2011/09/28/28climatewire-chinas-city-of-the-future-rises-on-a-wastela-76934.html?pagewanted=all Accessed 1 March 2014.

53 Royal Ten Cate (2013) Ten Cate Geotube® wastewater impoundment lake remediation – Tianjin Eco-City, China. Available at: http://www.tencate. com/apac/geosynthetics/case-studies/dewater-tech/news-dewat3.aspx Accessed 1 March 2014.

54 Royal Ten Cate (2013) *Ten Cate Annual Report*. Almelo, Royal Ten Cate, 58.

55 Baeumler, A., Chen, M., Dastur, A., Zhang, Y., Filewood, R., Al-Jamal, K., Peterson, C., Randale, M. and Pinnoi, N. (2009) *Sino-Singapore Tianjin Eco-City: A Case Study of an Emerging Eco-City in China*. Washington, DC: The World Bank.

56 Liu, W., Lund, H., Mathiesen, B. V. and Zhang, X. (2011) Potential of renewable energy systems in China. *Applied Energy* 88(2): 518–25.

57 *China Daily* (2009) China eyes 20% renewable energy by 2020. *China Daily*, 10 June 2009. Available at: http://www.chinadaily.com.cn/china/2009-06/10/ content_8268871.htm Accessed 1 March 2014.

58 Li, Y. and Currie, J. (2011) *Green Buildings in China: Conception, Codes and Certification*. Washington, DC: Johnson Controls.

59 Chu, A. and Chan, J. (2013) *Cleantech in China: Building a Green Future*. London: PricewaterhouseCoopers.

60 Boland, A. (2007) The trickle-down effect: ideology and the development of premium water networks in China's cities. *International Journal of Urban and Regional Research* 31(1): 21–-40, 30.

61 Pow, C.-P. (2009) *Gated Communities in China: Class, Privilege and the Moral Politics of the Good Life*. London, Routledge.

62 Hellström Reimer, M. (2010) Unsettling eco-scapes: aesthetic performances for sustainable futures. *Journal of Landscape Architecture* 5(1): 24–37; Wong, T. C. (2011) 'Eco-cities in China: pearls in the sea of degrading urban

DOI: 10.1057/9781137298768.0005

environments.' In Wong, T.-C. and Yuen, B. (eds) *Eco-City Planning: Policies, Practice and Design*. New York, Springer: 131–50.

63   Springer, C. (2012) Public housing in the Sino-Singapore Tianjin Eco-City: the (missing) link between social cohesion and green living. *EcoCity Notes*, May 2012. Available at: http://ecocitynotes.com/features/eco-city-demographics/public-housing-in-the-sino-singapore-tianjin-eco-city/ Accessed 1 March 2014.

64   Ho, A. L. (2013) Growing pains for Tianjin Eco-City. *The Straits Times Asia Report*, 13 October 2013. Available at: http://www.straitstimes.com/the-big-story/asia-report/china/story/growing-pains-tianjin-eco-city-20131013 Accessed 1 March 2014.

65   Springer, *Public housing in the Sino-Singapore Tianjin Eco-City*, 2.

66   Li, Y. (2012) Environmental state in transformation: the emergence of low-carbon development in urban China. In Holt, W. G. (ed.) *Urban Areas and Global Climate Change*. Bingley, Emerald: 221–46, 237.

67   Baeumler *et al.*, *Sino-Singapore Tianjin Eco-City*.

68   Chang, C. I-C. and Sheppard, E. (2013) China's eco-cities as variegated urban sustainability: Dongtan eco-city and Chongming eco-island. *Journal of Urban Technology* 20(1): 57–75.

69   Rapoport, E. (2014) Utopian visions and real estate dreams: the eco-city past, present and future. *Geography Compass* 8(2): 137–49.

70   China Labour Bulletin (2014) Migrant worker wages increased by 14 percent in 2013. China Labour Bulletin website, 21 February 2014. Available at: http://www.clb.org.hk/en/content/migrant-worker-wages-increased-14-percent-2013 Accessed 3 September 2014. For 2010 figures, see China Daily (2011) China's 2011 average salaries revealed. *China Daily* website, 6 July 2012. Available at: http://www.chinadaily.com.cn/china/2012-07/06/content_15555503.htm Accessed 1 March 2014. For migrants' average salary see Fang, C., Wang, M. and Yang, D. (2013) Understanding changing trends in Chinese wages. In Huang, Y. and Yu, M. (eds) *China's New Role in the World Economy*. London, Routledge: 69–88.

71   Abramson, D. B. (2008) Haussmann and Le Corbusier in China: land control and the design of streets in urban redevelopment. *Journal of Urban Design* 13(2): 231–56.

72   Hodson, M. and Marvin, S. (2010) Urbanism in the Anthropocene: ecological urbanism or premium ecological enclaves? *City* 14(3): 365–9.

73   Ren, X. (2013) *Urban China*. Cambridge, Polity Press, xix.

74   Ren, *Urban China*, xix–xx.

75   Smil, V. (1993) *China's Environmental Crisis: An Enquiry into the Limits of National Development*. New York, M. E. Sharpe, 145.

DOI: 10.1057/9781137298768.0005

3

Peak Oil and Eco-Urbanism in Abu Dhabi

Abstract: *Abu Dhabi, in the United Arab Emirates, is built on oil wealth and relies on an economy which is largely reliant on fossil fuels. The chapter situates Abu Dhabi's development within the 'Age of Crisis' by focusing on key concerns facing the emirate's planners and policymakers: these include the Peak Oil scenario, the need for economic transition and diversification away from extractive industries and towards a high-tech economy, and a rapidly changing demographic profile as a result of migratory inflows. The chapter investigates the example of Masdar City, a high-tech eco-city being built near the emirate's capital, and asks critical questions about the meaning of sustainability in this eco-city mega-project.*

Caprotti, Federico. *Eco-Cities and the Transition to Low Carbon Economies*. Basingstoke: Palgrave Macmillan, 2015. DOI: 10.1057/9781137298768.0006.

Hydrocarbons and urban development

Abu Dhabi's steel and glass skyline speaks volumes about the ascent of the emirate of the same name in the mid- to late-20th century, and especially in the last few decades. Camel herding, date production, fishing and pearl diving were the emirate's economic mainstays up to the mid-20th century. Governance was mainly concentrated in the hand of sheikhs belonging to the dominant Bani Yas tribe, a branch of which controlled the emirate of Dubai to the south-east.

Abu Dhabi is now known as one of the federated emirates of the United Arab Emirates (UAE). The UAE is a federation of seven emirates which gained independence in 1971: Abu Dhabi is their capital, as well as the largest state. Although Dubai, its glittering neighbour, is perhaps the most widely known symbol of the rise to global prominence of the UAE, Abu Dhabi is in actuality the wealthiest of the emirates in terms of both GDP and income per capita. Yet, until the early 1960s, Abu Dhabi was not particularly wealthy, although it was geopolitically important to Great Britain, which protected it (and the other emirates) because of its strategic position near key shipping and trade routes.

The discovery (in the 1950s) and exploitation (from 1962) of oil in Abu Dhabi radically changed the emirate. The Emirate's first paved road was built in 1961. By 1966, the capital city's roads were 'paved,' but only with gatch, a local limestone material also known as hardpan. There were few cars to use the roads anyway: by 1966, only 30 cars were registered as being in circulation in the capital of the UAE.[1] The few concrete buildings built in the city in the 1950s were a maximum of five floors high. The urban and economic geography of Abu Dhabi was, in the 1960s, just starting to change: its development accelerated as the oil industry grew.

Oil revenues grew as production increased and were partly reinvested in developing Abu Dhabi. Oil production has increased enormously in the emirate, from 120,000 barrels a day in 1963 to 1.6 million barrels per day in 2014.[2] There are plans to increase this production level to 3.5 million barrels per day by 2017.[3] Revenues have increased apace: from just under US $2 million in 1962, to US $190 million by the end of that decade, to US $71.2 billion in 2007.[4] The implications for Abu Dhabi of its nascent oil industry were clearly enormous; the emirate was transformed from an economy which was not reliant on large-scale infrastructure or extractive industries, to one which was literally 'plugged in' to (and dependent on) the global energy landscape. Levels of per capita GDP

DOI: 10.1057/9781137298768.0006

grew rapidly. In 1970, just eight years after commercial production of oil started in the emirate, GDP per capita was of around US $13,340. By 2010, this had risen to US $88,483.[5] In the 2010s, GDP per capita has consistently been above the US $100,000 per capita mark, making Abu Dhabi one of the wealthiest parts of the world.

Accompanying the production and export of oil has been the development of infrastructure, from pipelines to access roads, to refineries and export terminals, which have enabled Abu Dhabi's oil to be sold on the global market. From the start of oil production to the present, over 20,000 tankers have sailed from the Jebel Dhanna oil export terminal with their cargo of black gold: in return, capital has flowed in to the emirate, first to the Abu Dhabi Company for Onshore Oil Operations (ADCO) and then, from January 2014, to the Abu Dhabi National Oil Company (ADNOC).[6] The emirate has also been active in oil refining since the construction of its first refinery, at Umm al-Nar, in 1976: it currently has a refining capacity of 90,000 barrels a day and the original refinery has since been joined by other refineries.[7]

The emirate's development has largely been reflected in its growth and in that of its capital city. Abu Dhabi had a population of 58,000 in 1952: this increased to over 2.3 million by 2012. By 2013, the city of Abu Dhabi was populated by almost a million residents; the government expects that the city's population will number around three million by 2030.[8] Abu Dhabi's contemporary nature is deeply urban, since over 80 per cent of the emirate's population lives in Abu Dhabi city itself. The city's growth can be seen in its new skyline and in the network of highways and bridges that connect its various neighbourhoods. Although the city's high-rise development has not been as headline grabbing as that of Dubai, there are still hundreds of high-rise buildings in the urban area, and many are new skyscrapers, such as the new 88-floor World Trade Center Abu Dhabi – The Residences, completed in 2014.

On a wider economic scale, the city has become a national and regional centre of economic importance. It houses key institutions such as the UAE's central bank and the Abu Dhabi Securities Exchange, and it is linked with international business and tourism centres through a rapidly expanding international airport. This growth in urban and national infrastructural, architectural and economic development has come as a direct result of the emirate's oil economy, and as a consequence of the investment of oil revenues for national development.

DOI: 10.1057/9781137298768.0006

Part of the reason for the emirate's success in utilising oil revenues for national economic and urban development projects has been the use of sovereign wealth funds as vehicles through which to invest oil income. In the case of Abu Dhabi, several large sovereign wealth vehicles have enabled the emirate to pursue a diversified range of investments. The largest sovereign wealth fund in the world in 2012 was the Abu Dhabi Investment Authority, which owned assets worth US $627 billion. The emirate's other sovereign wealth funds include, among others, the International Petroleum Investment Company, which manages assets worth US $65 billion, and the Mubadala Development Company (MDC), whose assets total US $48 billion.[9] All of these funds invest earnings derived from oil revenue.[10] Abu Dhabi is therefore well-placed to direct capital flows strategically into urban and other forms of development, although this strategy has not come without its problems.

It is clear that Abu Dhabi has known a rapid economic growth curve since the discovery and exploitation of onshore oil in the middle of the 20th century. Hydrocarbon wealth has powered Abu Dhabi's meteoric rise from a regional backwater to a nation of world importance, and has fuelled the emirate's urbanisation. This can be seen in the glass and steel skyline of Abu Dhabi, as mentioned above. It is also symbolised by several recent landmark developments such as the Yas Marina Formula 1 circuit, or the establishment of flagship yet franchised cultural institutions, such as the Guggenheim Abu Dhabi and the Louvre Abu Dhabi. These urban projects have rapidly become global signifiers aimed at placing both the city and the emirate firmly on the symbolic map of global cities. The emirate's accumulation of large surpluses can also be seen in the appearance of the world's first gold bar vending machine, in the lobby of the Emirates Palace hotel.

Abu Dhabi's significance in the regional and global oil economy is without question: the emirate owns around eight per cent of global oil reserves, and its sovereign wealth has enabled it to take stakes in well-known international corporations and ventures overseas.[11] As Gulf scholar Christopher Davidson has argued in his book, *Abu Dhabi: Oil and Beyond*:

> With ninety years of remaining hydrocarbon production and with plans to increase oil output by 30 per cent in the near future Abu Dhabi will have the resources and surpluses it needs – regardless of the vagaries of broader economic trends – to extend considerably its historic strategy of building up petrodollar-financed overseas investments. With acquisitions across Asia,

DOI: 10.1057/9781137298768.0006

Africa, and increasingly in Western Europe and North America – including a major stake in Citigroup and ownership of the iconic New York Chrysler Building – the emirate's plethora of government-backed investment vehicles already control funds several times greater than those of Kuwait, China, Norway, and other prominent asset-managing states.[12]

Spectacular experimental urbanism

Abu Dhabi's re-investment of oil revenues into a centrally directed national economic development plan has resulted in the shape and look of the emirate as witnessed today. In a sense, the whole of the UAE is an experimental urban project. This is because the rise of Dubai, Abu Dhabi, and the other emirates since the mid-20th century as a result of the emergence of a hydrocarbon economy in the Gulf has led to the potential for the country's leaders to envision specific 'futures' for their emirates, and to build these futures by sinking parts of their oil revenue streams into materialising these visions of the Gulf's future contours. Therefore, emirates such as Abu Dhabi can be seen as experimental from the beginning since their development was predicated not only on establishing pathways to an envisioned level of economic development, but also on developing systems of governance, sovereign investment, and urban growth which would enable the achievement of development targets. Furthermore, in light of the fact that development is centred on the UAE's main cities, Abu Dhabi's experimental character can also be described as quintessentially urban. The construction of new and spectacular urban projects can be seen as the latest evolution in a decades-long process of experimenting with the economy of emirates such as Abu Dhabi.

There is no doubt that much of the focus of urban development in the UAE and the wider region in the past two decades has been on spectacular urbanism. For example, the UAE and Saudi Arabia are rapidly becoming the focal points for the construction of ultra-tall buildings. Examples of the spectacularisation of architecture and of the urban sphere in hydrocarbon economies are the Burj Khalifa, an 828-metre skyscraper in Dubai, and the construction (underway at the time of writing) of the Kingdom Tower, a skyscraper that may reach over one kilometre in height. Other examples of spectacular urbanisation in the region include projects predicated on creating new urban environments,

DOI: 10.1057/9781137298768.0006

such as the Palm Jumeirah, an artificial archipelago of islands built on reclaimed land and resembling a palm, or Saadiyat Island in Abu Dhabi, another island based on reclaimed land and scheduled to house flagship projects such as the Louvre Abu Dhabi. Furthermore, the construction of world-class airport terminals throughout the UAE and the funding of national airlines (such as Dubai's Emirates or Abu Dhabi's Etihad) to turn the UAE into a global air travel hub are long-term initiatives aimed at firmly placing cities such as Dubai and Abu Dhabi on the world map, both materially and symbolically.

Analyses of spectacular urbanism have been prominent in recent years, and the UAE has understandably been a focus for scholars and commentators interested in the link between entrepreneurial urbanism, symbolism, and place-making in emergent cities. Much of this interest has focused on Dubai, Abu Dhabi's sister city, where symbolic and spectacular architecture coupled with the construction of 'vertical cities' has attracted much critical attention.[13] However, there has been significantly less analysis of the spectacular nature of urban development in Abu Dhabi, although several scholars have recently started working on the implications of urban planning in the political and socio-economic context specific to the emirate.[14] Nonetheless, while Abu Dhabi's urban development may not be populated by height record-beating skyscrapers, the emirate is replete with examples of more low-rise, spectacular, and entrepreneurial urban developments. Furthermore, it can be argued that Abu Dhabi's urban development plan is, in its totality, a plan aimed at spectacular urbanisation.

Abu Dhabi's urban development trajectory has, since the late 2000s, been focused on urban transition. The lack of coherent planning frameworks up to the 21st century had led to Abu Dhabi's rapid emergence as a city without a clear architectural and design vision for the future. However, the formation of a national urban planning body, the Urban Planning Council (UPC), in 2007, heralded the professionalisation of planning in the emirate. The UPC was given the remit of defining the emirate's urban shape, and of establishing best practice requirements for planning. In September 2007, the UPC produced the Abu Dhabi 2030 urban development plan, which enshrines the emirate's vision for the city's trajectory to 2030.[15] The plan is not simply a master plan for the city of Abu Dhabi, but for the whole emirate: it is a 'mega-plan' which focuses on the wider mega-project of refashioning the whole of Abu Dhabi in the image of global economic success.

DOI: 10.1057/9781137298768.0006

The UPC has become the institution responsible for envisioning the emirate's urban future. In so doing, it is not only an authority focused on transition towards a specific urban development target, but is also an example of the 'rule of experts' in defining urban and national futures.[16] This can be seen in the fact that the Abu Dhabi 2030 plan was developed by an elite group of policy, planning, and private sector professionals involving not only international planning 'experts' but also Emirati business leaders. The inclusion of non-expert views was largely absent in the elaboration of plans for the emirate's future, apart from some interviews which were carried out with local residents.

The involvement of urban planning and other elites included the Boston Consulting Group (who elaborated the initial assessment report prior to the Plan); planners from Vancouver's city planning department and Canadian urban design firm Perkins+Will; global engineering firm Arup (who advised on infrastructure); and Los Angeles-based Economic Research Associates (which developed the plan's economic growth targets), among others.[17] The Abu Dhabi 2030 plan originated some immediate positive impacts, such as the halting of a planned expressway through Abu Dhabi's downtown.[18] Nonetheless, what is clear is that the urban visions for Abu Dhabi's urban future, as enshrined in the UPC's plans, are both transitional and determined by local and international 'experts'. This has often caused a clash between visions of the urban environment as found in the marketing materials produced by architectural and planning consultancies, and the everyday, lived city. Yasser Elsheshtawy, an architecture scholar at UAE University in the emirate of Al Ain, has described the disjuncture that is often apparent between visions of spectacular urbanism on the ground, in urban locales where the spectacular city 'clashes' with the human experience of the urban:

> The necessary city ingredients are there – streets, high-rise buildings, land-marks – but there is a certain detachment between these urban symbols and the city's citizens as well as its surrounding barren, desert landscape, imparting a strong sense of artificiality.[19]

In spectacular eco-urban projects such as Abu Dhabi's Masdar eco-city, the key issue of what the link between elite visions of the city, and what the 'finished product' of the eco-city will be like for its inhabitants and for those working within the city remains an open question.

DOI: 10.1057/9781137298768.0006

Crisis and eco-urbanism in the UAE

The Age of Crisis has not spared the UAE or Abu Dhabi. The financial crisis of 2008 (which required Abu Dhabi to bail out neighbouring Dubai) was evidence that inflows of petrodollars cannot be considered the only source of security in the region. Abu Dhabi's oil-based economy is potentially more resilient than that of Dubai, which saw a housing and real-estate bubble in 2004–8 that led a crash in property prices of 50 per cent in 2009 alone.[20] It is interesting to note that key, iconic buildings such as the Burj Khalifa were inaugurated in the midst of the financial crisis. This underlines what has been argued by geographer Maria Kaika, that iconic buildings are not necessarily signifiers of economic success, but 'architectures of crisis', symptomatic of often traumatic economic and other changes.[21] Abu Dhabi's spectacular eco-urbanism can also be situated within this framework, interpreting the construction of iconic projects as expressions of deeper change and crisis.

There is clearly a set of 'crises' threatening Abu Dhabi's socio-economic future. Many of these crises are felt across the region, albeit in different configurations. For example, one of the key environmental-economic issues facing the emirate is the Peak Oil scenario. Although there are strong doubts as to whether Abu Dhabi has passed its oil production peak, it is increasingly clear that the emirate cannot rely on oil revenues indefinitely. Abu Dhabi's current success is clearly founded on oil wealth: the emirate owns 95 per cent of the UAE's oil resources.[22] As a result of the effective development of its oil economy, Abu Dhabi's economic growth has been significant, achieving a GDP growth rate of eight per cent in 2010.[23]

Recognition of the need to diversify Abu Dhabi's economy (and especially its energy mix) away from an overwhelming dependence on the hydrocarbon economy has led the emirate's government to consider economic development strategies that are not based on the oil sector. The key idea is not to stop producing oil, but rather to realise even higher earnings from oil exports (as oil prices are expected to rise with increasing demand) when resources start to dwindle, while enabling Abu Dhabi to wean itself off oil as its main energy source. At the same time, the emirate's transitional economic development trajectory is clearly being fashioned so as to diversify away from the oil sector. The concern with viewing oil as an *export* rather than as a resource to be used for national energy production is also a key plank of development trajectories for

DOI: 10.1057/9781137298768.0006

other states in the region. Saudi Arabia, for example, recently announced its intent to continue exporting oil while becoming 100 per cent reliant on renewable and nuclear energy for national energy consumption.[24]

Therefore, while Abu Dhabi's fossil fuel reserves are considerable in extent, there has been a recent, urgent sense of a need to diversify the economy away from its primary reliance on oil and gas. This is in part due to the emirate's large carbon footprint, but is also largely due to the recognition that continued, rapid economic and population growth will need to be fuelled by an energy mix that is diverse. This has led to Abu Dhabi setting a national renewable energy target of seven per cent of national energy generation to be derived from renewables by 2020. This target is far from impressive: a recent study has highlighted the fact that out of 61 countries with renewable energy generation targets, the UAE was the only one with a target below 15 per cent by 2020.[25]

The transitional impetus which characterises much of Abu Dhabi's economic and urban policy initiatives and projects can be seen at a macro scale when considering the continuing diversification of the economy. For example, in 2010, 70 per cent of Abu Dhabi's GDP was directly derived from hydrocarbon revenues.[26] By the end of 2011, and in great part as a result of strategies of economic diversification and transition away from over-dependence on oil, that figure had decreased to 58.5 per cent.[27] The gradual shift from an oil and gas economy towards a more diversified energy economy has been described as part and parcel of Abu Dhabi's wider attempt to rebrand itself from the mid-2000s onwards.[28]

The construction of spectacular architecture, new neighbourhoods, and flagship museums and sporting facilities are paralleled by an economic rebranding which still includes oil, but which increasingly highlights new, gleaming technologies associated with alternative energy sources, and other energy infrastructures. In turn, Abu Dhabi's shift towards a more diversified energy-based economy is part of a specific transitional strategy, one that is largely top-down centrally and strategically directed by political, economic, and technical elites. In the case of Abu Dhabi, the dual challenges of how to diversify the economy away from almost total reliance on oil, and how to diversify the national energy generation and consumption landscape are issues that will have significant economic and environmental consequences.

Nonetheless, there are also concerns about environmental 'crises' in Abu Dhabi, as the emirate develops and its population expands. The emirate's development in the past 60 years or thereabouts has

DOI: 10.1057/9781137298768.0006

clearly changed the ecology of a wide swathe of its marine and land environments. Urbanisation, land reclamation, oil production, and reliance on energy-intensive technologies (from desalination to air conditioning) for enabling and sustaining contemporary standards of living have all contributed to the emirate's significant demand on its natural resources. With regards to urban growth, for example, concerns as to the sustainability of Abu Dhabi's urban expansion are underlined by the fact that the ecological footprint of the city of Abu Dhabi is currently the largest in the world. Abu Dhabi's Environment Agency (EAD) calculates that each of the city's residents need to be sustained by 10.3 global hectares of land.[29] By comparison, Vancouver residents require 7.71 hectares per capita.[30] Furthermore, the emirate's urban growth in the context of a hot desert climate has contributed to consistently high energy demand. In the summer, 75 per cent of energy consumed in Abu Dhabi is used to power cooling and air conditioning systems.[31]

In the context of climate crisis, it is important to note that Abu Dhabi is especially dependent on coastal areas which will be the most affected in a scenario of rising sea levels as predicted by the IPCC. The emirate's vulnerability to climate change is stark: over 85 per cent of Abu Dhabi's population and 90 per cent of its industrial infrastructure exist within a narrow strip of land next to the coast. In 2009, in preparation for the Copenhagen climate conference, the EAD estimated future impact scenarios for rising sea levels in the Gulf: a sea level rise of 50 centimetres is projected to cause significant sea encroachment and the destruction of up to US $400 million in real estate. A sea rise of a metre (the IPCC expects a rise of 50 centimetres to a metre by 2100) or more will cause the flooding of 722 square kilometres of land in the emirate, impacting heavily on its capital city.[32] This is why some of Abu Dhabi's new urban projects, such as Saadiyat Island, have required enormous earthworks to raise the island and construct breakwaters against storm surges and rising seas. And yet, in the context of a shifting marine environment, it is nonetheless interesting that much of the spectacular development currently underway in the emirate is at the ocean's edge, or on reclaimed islands. The use of islands throughout the UAE as 'anchors' for place branding has been called 'Islomania' by Pernilla Ouis, a social scientist at Malmö University, Sweden:

> The country is struggling to transform its economy from fossil fuel export into tourism and finance. Islomania is one strategy to create attractiveness

DOI: 10.1057/9781137298768.0006

in a rather barren environment, somehow empty both in terms of natural wonders and historical sites. The new islands fulfill many needs in Emirati society: as new urban flagships necessary for economic diversification, as icons of the traditional past, and as places enabling economic, religious, social and cultural segregation. A flagship denotes spectacular grand buildings or other centers to stimulate urban development and attract investors, tourists and residents on the global market.[33]

Finally, one of the key issues facing Abu Dhabi's future is not related to oil, gas, infrastructure, sea level rise, or the changing climate. Rather, it is linked to social and demographic change in the context of the quick and wide-ranging transitions which have impacted the emirate in the past few decades, and which seem to be accelerating in terms of the speed of change. The wider demographic composition of Abu Dhabi has been in constant flux for decades: its population doubled between 1986 and 2005, mostly due to the influx of temporary workers. The Abu Dhabi 2030 plan is based on an estimate of a further doubling of the population to three million by 2030.[34] Clearly, the management of this wide-scale process of demographic growth is a necessity. At the same time, it introduces questions as to the role of national, ethnic, religious, and other forms of identity in Abu Dhabi, and in particular on the ways in which belonging to a specific socio-economic group may have a significant impact on opportunities, belonging, and rights.

Abu Dhabi's population is highly stratified in socio-economic terms. Only 25 per cent of its residents can call themselves nationals of the UAE. The other 75 per cent are expatriate temporary workers, of which only one per cent are high-income individuals.[35] This presents a picture of an economy which is largely performed by non-national workers of various nationalities and ethnicities, but who have vastly different rights of access to public and other services. This state of affairs naturally leads to questions around the stability and social sustainability of a society in which the residency status of most of the population is *de facto* precarious, and in which regulations promoting nationals – such as the much-vaunted 'Emiratisation' policy – are clearly and preferentially aimed at a small part of the overall population. Variants of the policy of Emiratisation can be found across the Gulf and in neighbouring states such as Saudi Arabia. Their impact is widely felt, especially at the higher end of the socio-economic scale: on 4 July 2013, Abu Dhabi's General Secretariat of the Executive Council, the highest policy-making body in the emirate, fired nearly its entire foreign staff.[36] Nonetheless, important questions remain

DOI: 10.1057/9781137298768.0006

as to how the low-paid workers on whose backs and sweat the emirate is being built will be integrated into the society of Abu Dhabi: this concern will be touched upon again in the conclusion to this chapter.

One of the key ways in which Abu Dhabi's emergence over the past few decades can be visualised is through its rapid urbanisation. It is no surprise, then, that building new urban environments is a key component of the emirate's attempts to engineer and think its way out of the crises and challenges identified above. Since the mid-2000s, a significant amount of capital and engineering and technical know-how have coalesced on a single, new-build eco-urban project which is being hailed as both an experimental urban centre, and as the city of the future: Masdar eco-city. This new city, inland from Abu Dhabi, is being built as a riposte to environmental concerns and is seen as a way of experimenting with a new urban environment which is more resilient in terms of energy production and consumption.[37] At the same time, Masdar eco-city is transitional in economic terms as well: as with Tianjin eco-city, Masdar is planned as a central node for the development of a high-tech economy focused on green and renewable energy technologies. The aim is to use a new city such as Masdar as a protected niche for the development of new technologies and a new economy, as the initial engine of economic-technological transition, and as a model urban environment for the future. Furthermore, the involvement of high-profile corporate partners at all stages of the eco-city's development also highlights the fact that one of the key features of the eco-city is its entrepreneurial nature: Masdar eco-city may be state-led, but it is being developed and built in close collaboration with private sector firms, and is expected to be one of the interfaces between Abu Dhabi and the global technology market. It is, in the words of geographer Federico Cugurullo, an example of the 'business of utopia' that increasingly links urban and economic development, sustainability, and visions of the 'good city' in new and at times uneasy configurations.[38] The rest of the chapter considers the case of Masdar eco-city in more detail.

Masdar eco-city: entrepreneurialism and the eco-city in Abu Dhabi

Masdar eco-city is being built about 17 kilometres south-east of Abu Dhabi, in a desert environment. The ground-breaking ceremony took

DOI: 10.1057/9781137298768.0006

place on 9 February 2008, and construction of the eco-city has progressed since, albeit not as quickly as was perhaps initially expected.[39] The city represents a very substantial investment by the Abu Dhabi government: the initial projected cost of designing and constructing Masdar was around US $15 billion.[40] By 2010, the total project cost stood at US $22 billion, and independent estimates placed the potential final cost at up to US $30 billion.[41] Most of the investment capital is sourced from the Mubadala Development Company (MDC), a multi-billion dollar government-owned investment firm.

By 2025, the city is expected to house 40,000 residents, and to feature corporate and other places of work for up to 50,000 commuters. At the time of writing, several buildings and streets have been built in and around the eco-city, as well as solar power plants and other infrastructural elements that are integral for the functioning of this urban project predicated on renewable energy technologies. Several of the buildings that have been constructed in the eco-city effectively demonstrate a commitment to building an urban environment which stands in stark contrast to that of Abu Dhabi proper. For example, Masdar eco-city currently features a tall 'wind tower' (*barjeel*), based on similar (albeit smaller in size) architectural elements common in cities in the Gulf. The wind tower provides a breeze at street level and, with its 45 metre height, is a powerful symbol of the attempt to fashion a new thermal eco-urban landscape in Abu Dhabi. In addition to the passive channelling of air down to street level, the wind tower also incorporates smart technologies such as sensors which enable openings at the top of the tower to open in the direction of prevailing winds, thus enabling the structure to capture airflow more efficiently. Furthermore, the tower incorporates mist generators, which add to its cooling effect.

The plan for Masdar city envisions a compact urban realm focused on low-rise buildings (a maximum of five floors is projected), narrow streets with a focus on pedestrians, and the intensive use of passive design elements (maximising shade and airflow, for example) while using innovative materials and technologies to ensure the city's sustainability in terms of energy.[42] Therefore, Masdar eco-city's wind tower is a material symbol of the ideals incorporated in the wider master plan for Masdar eco-city.

The Masdar master plan was developed by London-based global architectural and design firm Foster + Partners, while the city's infrastructure was designed by Mott MacDonald, a UK-headquartered

DOI: 10.1057/9781137298768.0006

engineering corporation. Thus, while the eco-city reflects elements of planning related to those found in other cities in the Gulf (such as the wind tower), it can also be described as a high-tech, normative vision of how an eco-city should function in the Gulf. At the same time, Masdar is clearly focused on the market and on being an urban pivot for transitional economic strategies. Underlying the use of innovative and cutting-edge technologies and the use of global brands and referents in the eco-city project, however, is also a specific view of what eco-urban planning is with regards to the sustainable city: a high-level endeavour developed largely by 'expert' actors in conjunction with the government of Abu Dhabi.

Masdar eco-city as an entrepreneurial eco-city

Masdar eco-city is an urban project that is central to Abu Dhabi's vision for its economic future. Nonetheless, one of the key characteristics of the project is that the eco-city is but one of the various strands which have been grouped together under the Masdar name. Indeed, 'Masdar' acts as a brand name for a series of concerns focused on clean technologies and associated markets, and is part and parcel of the wholesale process of imagining Abu Dhabi as a 'green' city through the construction of urban mega-projects such as the eco-city.[43] In this sense, Masdar eco-city is deeply entrepreneurial: its construction within a wider market-focused, transitional strategy which places investment, capital and potential returns, and economic performance at the very heart of a transitional vision for Abu Dhabi's environmental-economic future.

The Masdar eco-city project initially fell under the remit of the Abu Dhabi Future Energy Company (ADFEC), which was founded in 2006 and was wholly owned by the MDC. ADFEC's aim was of aiding Abu Dhabi in economic transition towards becoming a key player on the world stage in the development and commercialisation of clean technologies ('cleantech'). Although Masdar and ADFEC became synonymous by the 2010s, Masdar eco-city was initially just one among four of ADFEC's core activities. These also included Masdar Carbon, Masdar Power, and Masdar Capital, which administers the Masdar Clean Tech Funds.[44]

In 2010, the Clean Tech Funds comprised over US $250 million in commitments from ADFEC, Consensus Business Group (CBG), Credit Suisse, and Siemens.[45] Masdar Capital has also seen investments from other sources, including Deutsche Bank's now-closed environmental investment branch, previously known as DB Climate Change Advisers:

DOI: 10.1057/9781137298768.0006

a joint fund was set up totalling US $290 million in investments, and named DB Masdar Clean Tech Fund. These amounts of capital can be seen as representing a *re-investment*: re-investing oil revenues into low-carbon technologies and technical solutions to the crises identified above. In this sense, the rapid emergence of government investment vehicles, such as Masdar Capital, which invest in low-carbon futures, is part and parcel of the wider trend which has seen the emergence of new sectors of the global economy, such as the cleantech sector, focused on the interface between climate crisis and technological solutions.[46]

Masdar eco-city is also entrepreneurial in other respects. One of the key drivers of the project has been the focus on envisioning the eco-city as the centre of a knowledge economy based on cleantech and on new, green technologies. In order to achieve this, a knowledge and research base was seen as a key requirement for the eco-city and for those who would live and work there. Therefore, one of the key initial investments made by ADFEC was the establishment, in February 2007, of the Masdar Institute of Science and Technology (MIST). The Institute is a graduate institution partnered with the Massachusetts Institute of Technology (MIT). It was billed as the first graduate institute that focused wholly on research into alternative energies and cleantech. Its centrality to visions for the eco-city can be seen in the fact that MIST was built on the eco-city site proper: it admitted its first students in September 2009. Thus, MIST is both part of the eco-city's master plan, and an integral part of its wider transitional mission, by contributing to the eco-city's cleantech R&D focus.

The transitional eco-city

Masdar eco-city can be seen as an expression of urban entrepreneurialism due to the involvement of corporate and market-based actors in envisioning, building, and operating the new city. However, the eco-city also needs to be seen in the wider context of the Abu Dhabi 2030 plan: Masdar is part of the national transitional plan to diversify Abu Dhabi, turning it into a global financial and high-technology development centre. In this light, Masdar eco-city is a central node around which this broader vision turns, because it is in Masdar eco-city where some of the key transitional technologies and economic principles on which the 2030 plan is based are to be tested: from solar power, to the promotion of renewable energy and other technology businesses in and near the eco-city (and through Masdar-linked investments) to the establishment

DOI: 10.1057/9781137298768.0006

of educational facilities such as MIST, which aim to generate the knowledge capital required for the eco-city's socio-technical role.

At the most fundamental level, Masdar eco-city exists as a target of investment by the MDC. The eco-city is a recipient of capital flowing directly *from* Abu Dhabi's hydrocarbon economy, and *into* its vision for a future low-carbon economy. Masdar eco-city can then effectively be seen as an interface between the oil economy and green capitalism: its aim is to turn revenues from a finite resource into the market potential to exploit renewable sources of energy as well as other green technologies. Therefore, the role of players such as the MDC in the Masdar eco-city project can be conceptualised in transitional terms, as providers of flows of capital aimed at economic-environmental transition. This can be clearly seen in MDC's mission statement, in which the company's aim is to catalyse 'the economic diversification of Abu Dhabi'.[47] As the director of Carbon Management at Masdar stated:

> Masdar aims to become a world-leading technology integrator, a company capable of delivering highly complex, technologically challenging, environmental solutions on a significant industrial scale. This strategy will deliver significant long-term economic and technological benefits, both to Masdar and Abu Dhabi. In the process of generating the technological and organisational solutions needed to deliver these projects, Masdar will develop and acquire the needed expertise and intellectual property. This human capital, Masdar's pool of talent and knowledge, will allow Masdar to grow and become a sophisticated driver of low carbon economic development in the UAE, the region and the world.[48]

In terms of economic transitional strategies, one of the approaches which are being put into place with the aim of positioning Masdar eco-city as a cleantech hub is the institution of a sizeable SEZ near the eco-city. The range of incentives offered to attract suitable participants to the low-carbon economy around Masdar eco-city includes zero import taxes and no taxes on corporations or individuals within the SEZ. The Masdar SEZ also has no restrictions on the movement of capital, the ability for firms to be wholly owned by non-Emirati individuals or corporations, and zero currency restrictions. Global talent is attracted through the use of economic and lifestyle incentives, as well as through institutions such as MIST. As the Masdar eco-city website claims:

> Masdar eco-city is emerging as a global centre for the sale, marketing, servicing, and demonstration of renewable energy and sustainability technologies

[...]. The city will be home to numerous companies, from start-ups to Fortune 500 multinationals to the most cutting-edge technology firms, bringing these businesses close to fast-growing markets of the Middle East and Asia – hungry for clean technology and renewable-power products and services.[49]

Part of a wider, national strategy around establishing Masdar eco-city as a centre for renewable energy technology research and development is Abu Dhabi's focus on creating a cultural economy around the renewable energy sector.[50] Faced with the difficulty of rebranding an emirate previously known as almost completely dependent on the oil industry, parts of the Abu Dhabi 2030 plan include establishing events and institutions which place Abu Dhabi on the global renewable energy 'map' in terms of enabling relationships with global renewable energy elites, thus promoting a view of the emirate as a rising power in green energy and environmental technologies. This has included events such as hosting the World Future Energy Summit and institutions such as the International Renewable Energy Agency (IRENA).[51] For example, the World Future Energy Summit – described as the 'Davos of renewable energy' – was first hosted in 2008, and aims to bring together leaders in renewables, in a well-attended and publicised event which has Abu Dhabi as its stage. The summit quickly gained traction and international credibility, and by the early 2010s it was regularly attended by world leaders. In 2011, Ban Ki-moon, the UN Secretary General, was one of the highest profile attendees; a total of 26,000 individuals from 112 countries attended the summit that year. This included 35 official delegations, 200 speakers, and 600 corporate and other exhibitors. In 2012, high-profile leaders were again key attendees. Keynote speeches were given by Ki-moon, as well as Wen Jiabao, China's premier, and Hwang-sik Kim, the prime minister of South Korea.[52]

These efforts (new events and institutions) can be seen as promoting a low-carbon cultural economy around Masdar. This is because they enable, firstly, *relationality* to be enacted between renewable and environmental technology professionals and investors, with Abu Dhabi as the main interface where the executives and corporations can meet and interact. Secondly, Abu Dhabi's strategy enables the *performance* of a cultural economy around the low-carbon economy by establishing institutions that propagate and shape 'green' agendas through reports, conferences and other relational mechanisms. Therefore, Masdar eco-city can clearly be seen as part and parcel of a market-focused transitional strategy which envisions the eco-city as the central node in a wider shift in Abu Dhabi's economic landscape.

DOI: 10.1057/9781137298768.0006

People and technology

A significant part of the envisioning of Masdar eco-city has been the focus of the eco-city's plans on the potential for technology to deliver solutions to the issues faced by Abu Dhabi in energy, economic, and environmental terms. Marketing materials and plans for the city are replete with descriptions and celebrations of the eco-city's high-tech nature. The celebration of green technologies has the effect of depicting Masdar eco-city as a futuristic, clean, sleek, and comfortable city in which urban life can be lived in a seamless, frictionless way while benefiting from the advantages that can be derived from technological solutions to the problems of urban living. For example, the eco-city's plans include highly visible technological 'solutions' such as Personal Rapid Transport pods to enable journeys within Masdar without the use of a motorcar.

One of the characteristics of the Masdar eco-city project is the challenge of building a viable, low-carbon urban environment in which it is pleasant to live *despite* the hot and arid conditions of the desert in which the eco-city is being built. The eco-city is presented as a crowning achievement of technology and investment capital in successfully engineering a city so as to be a 'green island' in the midst of a desert environment seen as largely hostile and certainly as a place in which it is hard to live in comfort. This view of the eco-city is based on a representation and conceptualisation of the desert which posits it as a negative 'first nature,' which has to be transformed through technology into a green, pleasant, cool, and low-carbon urban 'second nature'. It is also at odds with cultural understandings of the desert which don't view it as necessarily hostile to human habitation. Nonetheless, in the Masdar eco-city project what is clear is (as was seen with the development of Tianjin eco-city on wetland) the fact that the new city is being constructed as a positive juxtaposition to an initial 'natural' state (the desert) conveniently described as hostile. Technology (in the case of Masdar eco-city, high-tech 'solutions' such as solar power), then, becomes constructed as the means through which positive urban environments can be delivered in a context of otherwise negative conditions. The eco-city becomes not simply a solution to Abu Dhabi's various current and future potential crises, but a fix for the 'necessity' of urbanising a desert environment, and an enabler for continued urban and economic growth.

The set of technologies that arguably lies at the heart of Masdar eco-city's identity are those technologies and infrastructural elements related

DOI: 10.1057/9781137298768.0006

to renewable energy, especially solar power. A significant amount of investment has been directed into planning Masdar eco-city so as to be powered exclusively by renewable energy. The city is engineered so as to make significant use of solar power, with the objective of making Masdar eco-city into a high-tech solar manufacturing and R&D hub. The aim of focusing the eco-city on solar power was in itself a transitional strategy: during the initial project development stage, Masdar eco-city relied heavily on overseas solar technology providers and manufacturers to provide the project with the required initial technology and installations. The utilisation of overseas ventures to provide an initial technology input was therefore part and parcel of a transitional process aimed at eventually enabling a solar power niche to be established in and around Masdar eco-city.

For example, an initial investment of US $2 billion was made in 2007 to finance the construction of a photovoltaic (PV) panel manufacturing plant in Erfurt, Germany.[53] At the same time, a 10 megawatt PV installation was completed near the Masdar eco-city site at a cost of US $50 million.[54] With regards to the construction of solar power generation facilities next to the Masdar eco-city site, Masdar reached an agreement with French oil company Total and Spanish solar firm Abengoa in June 2010 to construct a 100 megawatt concentrating solar power (CSP) facility named Shams-1.[55] This exemplifies the Masdar eco-city project's initial focus on enabling an Abu Dhabi-based solar industry to emerge, based on an initial strategic emphasis on imported PV and other technologies. The PV and CSP installations around the eco-city can therefore be seen as both powering the specific urban environment of the eco-city, and as useful experimental ways of promoting the inflow of technologies and technical expertise in light of Abu Dhabi's broader transitional aims of diversifying its energy mix away from the hydrocarbon economy.

Furthermore, the eco-city's master plan calls for a significant focus on green building and on the integration of new building technologies into Masdar's urban fabric. An example of this is the construction within the eco-city of the Siemens headquarters: the building incorporates a range of passive and active elements which enable its environmental footprint to be considerably reduced. The headquarters are being constructed to LEED-Platinum standard, and the fact that they are to serve as a regional headquarters for a major corporation like Siemens points to the central role envisioned for private sector actors in the Masdar eco-city project. It also points to the consistent focus on urban sustainability understood in

DOI: 10.1057/9781137298768.0006

terms of a sustainable urban physical environment, and of a city where flows (of energy, carbon, materials and the like) can be tracked, measured, and managed.

It is, therefore, clear that Masdar eco-city is a project that is largely driven by economic and technological concerns. The city's reason for being is largely as a container of corporations and high-tech, white-collar workers, who will form an interface with the global market and, in so doing, function as an efficient and (it is hoped) successful cleantech and renewable energy industry cluster in Abu Dhabi. The foundation of MIST, with its clear technical focus, and the location of Masdar eco-city in the context of a wider SEZ is indicative of a project which aims to place a technical knowledge economy at the heart of an urban area conceived as an experimental economic-technological niche.

The question remains, however: in a city celebrated (and rightly so) for its technologically innovative nature, and for its focus on stimulating the growth of a low-carbon economy in Abu Dhabi, what kind of *social* urban landscape is being built? To date, there seems to have been much less attention devoted to questions around the social sustainability and social resilience of the eco-city. As Cugurullo has argued, projects such as Masdar eco-city can be seen as expressions of a specific permutation of the ideology of sustainability, in which economic concerns eclipse the arena of social sustainability:

> Masdar eco-city is [a] non-place: a non-anthropological spatial entity bereft of an organic society. A space where identity is suspended and little or nothing is left for emotional relations. Is this what the eco-city phenomenon is all about?[56]

This may be, in part, because of the limitations inherent in contemporary flagship urban planning, based as it is on a focus on plans and designs focused on spectacular visions and buildings as 'containers' of socio-economic urban life. The concept and ideal of the eco-city, and of the sustainable city, more specifically, has in the case of Masdar eco-city been interpreted through this highly technological lens.

To conclude, it is useful to draw on what is perhaps one of the clearest and most incisive critical accounts of the development of Abu Dhabi's urban form: Michael Cameron Dempsey's *Castles in the Sand*. Dempsey, a planner who worked in a range of roles in Iraq, Afghanistan, and Abu Dhabi, presents a complex picture of the clash between urban development, artificial environments, the oil economy, and the mirage of sustainability.

DOI: 10.1057/9781137298768.0006

He is clear about the fact that what is being built in the sands of the emirate is a vision of the urban future which may have lasting impacts not just in the UAE, but much further afield: 'Abu Dhabi's ongoing experiment in city-building, in addition to being a captivating story in its own right, has profound implications for our increasingly urbanized planet.'[57] In Masdar eco-city, as in other eco-city mega-projects across the world, visions of a new but not unproblematic urban future, built as a response to crisis, are being trialled and solidified into actually existing city environments.

Notes

1 Ghazal, R. (2011) When Abu Dhabi had 30 cars. *The National*, 3 November 2011. Available at: http://www.thenational.ae/news/uae-news/heritage/when-abu-dhabi-had-30-cars Accessed 1 March 2014.

2 Hellyer, P. (2014) End of a 75-year era of oil-fuelled progress for Abu Dhabi. *The National*, 9 January 2014. Available at: http://www.thenational.ae/business/oil/end-of-a-75-year-era-of-oil-fuelled-progress-for-abu-dhabi#page1 Accessed 1 March 2014.

3 Matsumoto, T. (2013) Abu Dhabi energy policy: energy problems plaguing Abu Dhabi and their implications for Japan. *Institute of Energy Economics of Japan*, August 2013. Available at: http://eneken.ieej.or.jp/data/5115.pdf Accessed 1 March 2014.

4 For 1960s oil revenue data see El Mallakh, R. (1970) The challenge of affluence: Abu Dhabi. *Middle East Journal* 24(2): 135–46. For more recent revenue data see Oxford Business Group (2010) *The Report: Abu Dhabi 2010*. London, Oxford Business Group.

5 GDP per capita was calculated by using 2010 prices in United Arab Emirates Dirham (AED), converted to US dollars using a 1:0.27 exchange rate from 2014. See Abu Dhabi's historical GDP per capita figures in AED in Statistics Center Abu Dhabi (2010) *Abu Dhabi Over Half a Century*. Abu Dhabi, SCAD.

6 Oil is now also exported from the ADCO-operated Fujairah terminal (Hellyer, *End of a 75-year era*).

7 Butt, G. (2001) Oil and gas in the UAE. In Al Abed, I. and Hellyer, P. (eds) *United Arab Emirates: A New Perspective*. London, Trident Press: 231–48.

8 UAEinteract (2013) Abu Dhabi's population at 2.33M, with 475,000 Emiratis. UAEinteract, 9 October 2013. Available at: http://www.uaeinteract.com/docs/Abu_Dhabi's_population_at_2.33m,_with_475,000_Emiratis/57590.htm Accessed 1 March 2014.

9 Sovereign Wealth Fund Institute (2014) Mubadala Development Company Available at: http://www.swfinstitute.org/fund/mubadala.php Accessed 1 May 2014.

DOI: 10.1057/9781137298768.0006

10   Bernstein S., Lerner J., and Schoar, A. (2013) The investment strategies of sovereign wealth funds. *Journal of Economic Perspectives* 27(2): 219–38.

11   Davidson, C. (2009) *Abu Dhabi: Oil and Beyond*. New York, NY: Columbia University Press.

12   Davidson, *Abu Dhabi*, 1.

13   Acuto, M. (2010) High-rise Dubai urban entrepreneurialism and the technology of symbolic power. *Cities* 27(4): 272–84.

14   Ponzini, D. (2011) Large scale development projects and star architecture in the absence of democratic politics: the case of Abu Dhabi, UAE. *Cities* 28(3): 251–9.

15   UPC (2007) *Plan Abu Dhabi 2030*. Abu Dhabi: Abu Dhabi Urban Planning Council.

16   Mitchell, T. (2002) *Rule of Experts: Egypt, Techno-Politics, Modernity*. Berkeley, CA: University of California Press.

17   UPC, *Plan Abu Dhabi 2030*.

18   Khirfan, L. and Jaffer, Z. (2013) Sustainable urbanism in Abu Dhabi: transferring the Vancouver model. *Urban Affairs*. Available at: http://onlinelibrary.wiley.com/doi/10.1111/juaf.12050/full (Accessed 1 March 2014).

19   Elsheshtawy, Y. (2008) 'Cities of sand and fog: Abu Dhabi's arrival on the global scene.' In Elsheshtawy, Y. (ed.) *The Evolving Arab City: Tradition, Modernity & Urban Development*. London: Routledge, 258–304.

20   Teather, D. (2010) Bailed out and broke, Dubai opens the world's tallest building. *The Guardian*, 3 January 2010. Available at: http://www.theguardian.com/business/2010/jan/03/burj-dubai-worlds-tallest-building Accessed 1 March 2014.

21   Kaika, M. (2010) Architecture and crisis: re-inventing the icon, re-imag(in) ing London and re-branding the City. *Transactions of the Institute of British Geographers* 35(4): 453–74.

22   Reiche, D. (2010) Renewable energy policies in the Gulf countries: a case study of the carbon-neutral "Masdar eco-city" in Abu Dhabi. *Energy Policy* 38(1): 378–82.

23   Khirfan and Jaffer, *Sustainable urbanism in Abu Dhabi*.

24   Harvey, F. (2012) Saudi Arabia reveals plans to be powered entirely by renewable energy. *The Guardian*, 19 October 2012. Available at: http://www.theguardian.com/environment/2012/oct/19/saudi-arabia-renewable-energy-oil Accessed 1 March 2014.

25   Mehzer, T. Dawelbait, G. and Abbas, Z. (2012) Renewable energy policy options for Abu Dhabi: drivers and barriers. *Energy Policy* 42: 315–28.

26   Reiche, *Renewable energy policies in the Gulf*.

DOI: 10.1057/9781137298768.0006

27 Murray, M. (2013) Connecting wealth and growth through visionary planning: the case of Abu Dhabi 2030. *Planning Theory & Practice* 14(2): 278–82.

28 Sim, L-C. (2012) Re-branding Abu Dhabi: from oil giant to energy titan. *Place Branding and Public Diplomacy* 8: 83–98.

29 Environment Agency Abu Dhabi (2014) Growth of Abu Dhabi. *Environmental Atlas of Abu Dhabi Emirate*. Available at: https://www.environmentalatlas.ae/pathways/growthOfAbuDhabi Accessed 1 March 2014.

30 Wilson, J. and Anielski, M. (2005) *Ecological Footprints of Canadian Municipalities and Regions*. Edmonton, Federation of Canadian Municipalities.

31 Al-Sallal, K. A., Al-Rais, L. and Bin Dalmouk, M. B. (2012) Designing a sustainable house in the desert of Abu Dhabi. *Renewable Energy* 49: 80–4.

32 Oxford Business Group, *The Report*.

33 Ouis, P. (2011) 'And an island never cries': cultural and societal perspectives on the mega development of islands in the United Arab Emirates. In Badescu, V. and Cathcart, R. V. (eds) *Macro-Engineering Seawater in Unique Environments: Arid Lowlands and Water Bodies Rehabilitation*. Berlin, Springer: 60–75, 60.

34 Ponzini, *Large scale development projects*.

35 Mohammad, R. and Sidaway, J. (2013) Spectacular urbanization amidst variegated geographies of globalization: learning from Abu Dhabi's trajectory through the lives of South Asian men. *International Journal of Urban and Regional Research* 36: 606–27.

36 The Economist (2013) Sending the foreigners home. *The Economist*, 13 July 2013. Available at: http://www.economist.com/news/middle-east-and-africa/21581783-sacking-foreign-civil-servants-may-become-regional-trend-sending Accessed 1 March 2014.

37 Davidson, C. (2010) Abu Dhabi's global economy: integration and innovation. *Encounters* 1(2): 101–28.

38 Cugurullo, F. (2013) The business of utopia: Estidama and the road to the sustainable city. *Utopian Studies* 24(1): 66–88.

39 Masdar (2008) Abu Dhabi's Masdar Initiative breaks ground on carbon-neutral city of the future. Press release, 9 February 2008. Available at: http://www.masdar.ae/en/mediacenter/newsDesc.aspx?News_ID=40&MenuID=55&CatID=44 Accessed 1 May 2014.

40 Masdar (2008) Abu Dhabi commits US$15 billion to alternative energy, clean technology. Press release, 21 January 2008. Available at: http://www.masdar.ae/en/mediacenter/newsDesc.aspx?News_ID=42&MenuID=55&CatID=44 Accessed 1 May 2014.

DOI: 10.1057/9781137298768.0006

41   Heap, T. (2010) Masdar: Abu Dhabi's carbon-neutral city. *BBC News*,
       28 March 2010. Available at: http://news.bbc.co.uk/1/hi/world/middle_
       east/8586046.stm Accessed 1 May 2014.

42   Caprotti, F. and Romanowicz, J. (2013) Thermal eco-cities: green building
       and urban thermal metabolism. *International Journal of Urban and Regional
       Research* 37(6): 1949–67.

43   Jackson, M. S., and Della Dora, V. (2011) From landscaping to terraforming?
       Gulf mega-projects, cartographic visions and urban imaginaries. In: Agnew,
       J., Roca, Z., and P. Claval (eds) *Landscapes, Identities and Development*.
       Farnham, Ashgate: 95–113.

44   Masdar Carbon invests in Clean Development Mechanism (CDM) projects
       and large-scale carbon abatement schemes; Masdar Power aims to develop
       utility-scale renewable energy generation and distribution within the
       emirate; Masdar Venture Capital is an investment arm which, through the
       Masdar Clean Tech Funds, makes diversified and direct investments in clean
       technologies. See the Masdar website (2014).

45   Masdar Clean Tech Funds (2010). Home: Masdar Clean Tech Funds
       Available at: http://www.masdarctf.com Accessed 1 May 2014.

46   Caprotti, F. (2012) The cultural economy of cleantech: environmental
       discourse and the emergence of a new technology sector. *Transactions of the
       Institute of British Geographers* 37(3): 370–85.

47   MDC (2010) About Mubadala. Available at: http://www.mubadala.ae/en/
       category/about-mubadala Accessed 1 May 2014.

48   Nader, S. (2009) Paths to a low-carbon economy – the Masdar example.
       Energy Procedia 1.1, 3951–8, 3958.

49   Masdar eco-city (2010) Business. Available at: http://www.masdarcity.ae/en/
       index Accessed 1 May 2014.

50   Berndt, C. and Boeckler, M. (2009) Geographies of circulation and exchange:
       constructions of markets. *Progress in Human Geography* 33: 535–51.

51   Mahroum, S. and Alsaleh, Y. (2012) Place branding and place surrogacy: the
       making of the Masdar cluster in Abu Dhabi. Faculty & Research Working
       Paper 2012/130/IIPI. INSEAD Abu Dhabi Campus: Abu Dhabi.

52   World Future Energy Summit (2012) Exhibition & Summit 2012. Available
       at: http://www.worldfutureenergysummit.com/Portal/about-wfes/
       overview/2011-summit-and-exhibition.aspx Accessed 1 March 2014.

53   Masdar (2007) Abu Dhabi heats up the global solar market with $2 billion
       investment in photovoltaic manufacturing. Press release, 27 May 2007.
       Available at: http://www.masdar.ae/en/mediacenter/newsDesc.aspx?News_
       ID=85&MenuID=55&CatID=44 Accessed 1 May 2014.

54   Meinhold, B. (2009) Masdar breaks ground on largest solar plant in
       Middle East. *Inhabitat*, 21 January 2009. Available at: http://www.inhabitat.

DOI: 10.1057/9781137298768.0006

com/2009/01/21/masdar-begins-construction-on-10mw-solar-power-plant
Accessed 1 May 2014.

55 Masdar (2010) Masdar partners with total and Abengoa Solar. Press release,
20 June 2010. Available at: http://www.masdar.ae/en/mediaCenter/newsDesc.
aspx?News_ID=144&MenuID=0&CatID=0 Accessed 1 May 2014.

56 Cugurullo, F. (2013) How to build a sandcastle: an analysis of the genesis and
development of Masdar eco-city. *Journal of Urban Technology* 20(1): 23–37, 34–5.

57 Dempsey, M. C. (2014) *Castles in the Sand: A City Planner in Abu Dhabi.*
Jefferson, NC: McFarland & Company, 1.

DOI: 10.1057/9781137298768.0006

4
Conclusion: Re-thinking the Eco-City?

Abstract: *The chapter evaluates current trends in eco-urbanism and critically interrogates the issues which have emerged from analysis of the two case studies of Tianjin and Masdar eco-cities. It argues for the need to critically engage with constructed notions of crisis, and for a critical rethink of the city as mere marketplace for 'green' technologies, products and services. The chapter concludes by arguing for the need to consider social sustainability as a key part of transitional eco-city projects, and makes the case for proposing alternatives to current plans and proposals for eco-city projects.*

Caprotti, Federico. *Eco-Cities and the Transition to Low Carbon Economies.* Basingstoke: Palgrave Macmillan, 2015. DOI: 10.1057/9781137298768.0007.

DOI: 10.1057/9781137298768.0007

Smart and resilient: sustainable urbanism and the eco-city

The discussion of eco-cities in this book can be framed within wider, ongoing debates about sustainable urbanism and the development trajectory of the cities of the future. In broad terms, the focus on eco-cities can also be placed within an emergent interest in societal futures: envisioning what socio-technical and economic-environmental trajectories may characterise the next few decades, and identifying desired pathways for change. Clearly, the imagining of future scenarios is a deeply political question, as well as a technical and scientific one: *who* decides which pathways are ideal? What institutions, and with what kind of public involvement, are going to be tasked with producing detailed visions of societal futures, and in what ways are these visions going to be open to contestation and debate? Notwithstanding these pressing questions over the intensely political nature of urban and societal futures and future trajectories, it is nonetheless clear that, as environmental studies scholar Frans Berkhout has recently argued, the current interest in environmental and other forms of change is shifting from analysis of the past to an emphasis on the future (from future change scenarios, to future strategies).[1]

The debate over the future of the city in the context of wider environmental and other changes has led scholars and policymakers to propose a range of interventions in the urban sphere as a way of enabling an urban response to change. One of the key concepts that has emerged is that of the 'resilient city', introduced in the first chapter. Resilience has been defined in a range of ways, but is commonly understood to function as a useful metaphor for describing the capacity of systems (such as cities) to rebound to their initial state after a shock such as a natural disaster.[2] It is a contested concept, in no small part due to the fact that mainstream notions of resilience present a seemingly implicit, conservative resistance to ideas of change and adaptation. Nonetheless, an understanding of urban resilience is becoming seen as increasingly important in light of the need to consider the city as a set of imbricated systems (from infrastructure, to information, to health and social systems), and in the context of the increasing awareness of the complex impact of shocks to these multiple but interconnected urban systems. Nonetheless, resilience has been to some extent captured within discussions of the eco-city, when the concept is designated as a key component of eco-city projects and strategies.

DOI: 10.1057/9781137298768.0007

Another notion which has recently come to the fore in debates over future sustainable urbanism is that of the Smart City. The Smart City concept is most closely associated with the idea of the primacy of information and telecommunications technologies (from networked information systems, to sensors, to 'Smart' applications and materials such as smart glass) in regulating urban life.[3] The way in which Smart Cities are often discussed in highly technical and technocratic terms, and in ways which feed into but do not adequately interrogate dominant, neoliberal forms of urban govern-ance, has been roundly criticised.[4] Nonetheless, there are many, and at times discordant, definitions of what it actually means for a city to be 'smart'.[5] These range from highly specific, technology-focused conceptualisations of the Smart City, to classifications that are basically re-iterations of current eco-city definitions. In an example of the latter, Chwen Jeng Lim, founder of Studio 8, a UK sustainable urban planning, architecture and landscape practice, and Ed Liu, a partner at UK architecture firm Barnaby Gurning Architects, describe what they call the 'Smartcity' in these terms:

> A central component of the Smartcity is urban agriculture and the estab-lishment of an ecological symbiosis between nature and built form... The Smartcity postulates that the next and final stage of evolution can only be a circular economy that subsumes agriculture, energy and industry into co-dependency and self-perpetuation.[6]

In the above definition, it is clear that what is being discussed by using the term 'Smartcity' could just as easily fit into a discussion of the eco-city, especially when considering the fact that the idea of integrating the circular economy into visions of sustainable development is already being implemented at the level of eco-city strategies in China.

Therefore, notwithstanding the emergence of interest in new concepts such as 'smart' and 'resilient' cities that has permeated much of the schol-arly dialogue on sustainable urbanism from the late 2000s onwards, the debate on eco-cities is characterised by remarkable longevity. Indeed, the early focus on eco-cities in the 1970s has been followed by an expan-sion of interest in and discussion of eco-city definitions, plans, projects, and indicator and evaluation frameworks involving not just academics but scholars, activists, policymakers, architects, planners, engineers and others. Judging by the adoption of eco-city strategies at the national level in several countries, and by eco-city projects currently under construc-tion, debates over the shape of eco-cities of the future look more and more like debates over *the* city of the future.

DOI: 10.1057/9781137298768.0007

Urban parallels

The discussion of Masdar and Tianjin eco-cities in the previous two chapters highlighted several parallels between the two projects. Admittedly, the specific configurations of state-corporate involvement are different between Tianjin and Masdar: an international governmental joint-venture involving key corporate actors in the case of the former, and a project led by a single government and also involving corporate actors in project design and delivery in the case of the latter. However, both eco-cities are capital-intensive mega-projects, in which the state and the private sector are seen as jointly responsible for delivering low-carbon urban futures as specified by technical and scientific experts. Both projects (with their individual differences) can also be conceptualised as iterations of neoliberal urban environmental management. They are also clearly experimental in both design and technological make-up, and both eco-cities have a transitional focus, although the specific 'crises' which the new cities are being engineered to respond to are different in their own specific national contexts. Nonetheless, both Tianjin and Masdar eco-cities are experimental, transitional projects aimed at low-carbon transitions, broadly conceived.

Both Masdar and Tianjin eco-cities are entrepreneurial in nature. This can be seen not just in the fact that both form part of wider economic strategies, but also in the design of each eco-city as an urban node on which SEZs and a whole raft of green economy-focused incentives can be targeted. Both eco-cities have been planned as interfaces with the global market, and as centrepieces for a knowledge economy based on high-tech, low-carbon industries. In the case of Tianjin, the city's marketing is clearly aimed at attracting white-collar workers, while in the case of Masdar eco-city, the location of MIST at the heart of the city speaks volumes about the type of knowledge economy that the project is trying to create. Tianjin eco-city's cleantech park highlights a wider focus on the cleantech sector as the future economic basis for the city; Masdar's SEZ and attraction of businesses such as Siemens is, likewise, a pointer to the project's green economy aspirations.[7]

At the same time, both Masdar and Tianjin eco-cities are examples of very specific urban visions for the future. In both cases, eco-urbanism means high-tech urban planning and design. Both projects are characterised by an experimental view of the city: one that is based in large part on the idea of the city as a *tabula rasa* on which new technologies and

DOI: 10.1057/9781137298768.0007

infrastructures can be used to solve the urban, economic, environmental, and social 'ills' of the Age of Crisis. This has had the consequence of ensuring that these eco-cities will be, for all intents and purposes, solidifications of the urban visions of a specific range of actors who can adequately be described as elite. In this sense, both Masdar and Tianjin eco-cities are not very dissimilar from the utopian New Towns and modernist city visions of the 20th century, with their focus on urban planning and new building techniques as the technical and scientific means through which better urban environments could be fashioned.

An additional parallel can also be found in the way in which both eco-city projects are being funded. In the case of Tianjin eco-city, the governments of China and Singapore, together with selected corporations, are investing in the project. This can be seen as a strategy of re-investment of government revenues into urban and technological 'fixes' to specific crises. The same can be said of Masdar eco-city, built as it is on the basis of a strategic targeting of part of the flow of petrodollars that floods into Abu Dhabi and contributes to its hydrocarbon-based wealth. Oil revenues sunk in the sand of the desert will, it is hoped, help to generate a 'clean' oasis from which a low-carbon economy will spring. New cities become the solutions to the future and current crises generated through an economic system that makes the construction of these eco-cities possible in the first place.

In terms of China and Abu Dhabi's urban futures, plans for Tianjin and Masdar eco-cities also highlight specific visions of *who* these new urban environments are for. In both cases, the high-tech nature of the eco-cities, and their market-facing, economically transitional characteristics, point to the eco-city as a high-value island where white-collar workers can work, live, and play. The vast amounts of thinking and planning which have flowed into street layouts, green building design, and infrastructural blueprints have by and large not been paralleled by thinking of the eco-city as a socially inclusive, diverse environment. Both Masdar and Tianjin eco-cities, therefore, risk falling into the category of high-tech visions for an urban future that leaves little room for alternatives in terms both of urban development and of the contours of China and Abu Dhabi's urban future. In so doing, they also risk functioning as engines for increasing visible inequalities and segregation. Furthermore, in selecting a transitional trajectory and fitting the eco-city to it, the question remains of how these cities will develop if economic, technological, and other trends should follow different pathways from those envisaged by enlightened planners, engineers, and architects.

DOI: 10.1057/9781137298768.0007

At the level of national and urban governance, one of the key parallels between eco-urban projects such as Tianjin and Masdar is the fact that both eco-cities are found within national political contexts that are not democratic. Neither China nor Singapore can be termed open democracies. In Abu Dhabi, as in the wider UAE, there are no political parties, and ruling families share power across the UAE, although limited elections have been allowed at the local government level. Abu Dhabi's hydrocarbon resources have directly contributed to its great wealth in part because of the way in which governance is structured in the UAE. Indeed, the various emirates are linked together in what can be called a federation, and the government of each individual emirate is considered a local government. Thus, the UAE functions according to a federal system, but the constitution assigns specific policy and economic areas strictly to the local (emirate) level. Thus, oil and gas reserves are exclusively governed by the emirate to which they belong: since most hydrocarbon reserves in the UAE are located in Abu Dhabi, the emirate has been able to benefit from its right to exploit these resources at the local level.[8] Hence, Masdar eco-city is being built in a context whereby the organisation of natural resource governance enables a single emirate to benefit from oil and gas revenues, while a centrally controlled planning and urban development system allows for visions of the urban future to be rapidly developed and deployed.

At the same time, this is effectively a closed system, which excludes the 'public' (in itself a deeply contested term in the context of the UAE) from the process of envisioning new cities and ecological urban futures such as Masdar eco-city. While international planning corporations and renewable energy firms are invited to provide the technical and engineering framework on which the city is being built, the involvement of those living and working in Abu Dhabi in designing the 'city of the future' is notable for its absence.

Another key parallel between Masdar and Tianjin eco-cities is the significant involvement of migrant workers in the wider urban context. The second chapter outlined the issue of migrant workers in and around the Tianjin area. There are similarities between the role of migrants in the eco-urban visions in both China and Abu Dhabi. In terms of Abu Dhabi's demographic constitution, it is telling that the emirate has been described as an 'ethnocracy,' in which ruling families and a minority of nationals are at the helm of national economic, urban, and other types of development.[9] While spectacular urbanism in the Gulf may focus on high-rise, steel and glass visions of the entrepreneurial,

DOI: 10.1057/9781137298768.0007

corporate city, there are other dimensions to the urban projects currently underway in the region.[10] These dimensions include transnational flows of migrants, who constitute the majority of Abu Dhabi's population. Thus, when considering the envisioning and construction of eco-cities such as Masdar, with their spectacular and highly visible technologies – from gleaming solar panels to green buildings – it is also key to consider how these petrodollar-funded urban areas intersect with the trajectories of migrants from all sides of the socio-economic spectrum.

It is telling that, while Masdar eco-city is being built as a green city of the future, and Saadiyat Island (the 'Island of Happiness') is emerging as a cultural gem, the workers who are building these projects are housed in segregated, often carefully guarded 'workers' cities'. These cities keep them far away from the new urban landscape of Abu Dhabi or Masdar, while separating low-wage construction and infrastructure workers from the urban areas where opulent display and spectacular architecture vie for media and world attention. However, as recently highlighted in a report by Gulf Labor (a coalition of artists concerned about workers' rights and conditions in the Gulf), there are other types of new cities rising in Abu Dhabi and in the wider UAE: cities such as Mafraq Workers' City, Abu Dhabi, built for a population of over 20,000 workers.[11] Constructed to a high standard, it is nonetheless a segregated city, guarded by security personnel. In the case of the construction of the Saadiyat Island development, outlined in the third chapter, workers are often housed in Saadiyat Accommodation Village (SAV), a camp for 20,000 workers. Described as a 'model of best practice' by the Saadiyat Island organisation, it has been critiqued in less complimentary tones by Gulf Labor:

> In many respects the SAV is similar to other labor camps; it is temporary housing tied to a construction project, hosting temporary workers while they are employed on Saadiyat Island projects, where onsite supervisors are still addressed as 'camp bosses', and where a broader national-level context of indebted migrant labor remains in place.[12]

Therefore, the key question remains of whether new eco-urban projects can really be called 'sustainable' if the only type of sustainability they really aim for is environmental-economic. What of *social* sustainability? The construction of mixed and open eco-cities, a far cry from eco-enclaves, surely seems a positive goal for urban planners, architects, and policymakers. As writer and designer Ann Lui has argued in her analysis of the building of the Guggenheim Abu Dhabi on Saadiyat Island,

DOI: 10.1057/9781137298768.0007

plans for buildings which are expected to be constructed according to the highest levels of environmental performance (buildings such as the LEED Platinum offices on the Masdar eco-city site) rarely if ever contain any element of planning for social sustainability or workers' rights:

> LEED, which has institutionalized metrics for the ever-nebulous goal of sustainability, includes provisions for building occupant comfort, yet barely mentions worker safety. Across the board, green projects have been no worse – but no better than – traditional projects in terms of worker safety... [T]he (lack of) 'sustainability' of forced labor by migrant workers remains to be folded into the accreditation process.[13]

Finally, one of the most striking parallels between Tianjin and Masdar eco-cities lies not in plans for the eco-cities themselves, but in the way in which these eco-cities are being positioned vis-à-vis *nature*. Both eco-cities are being built in environments which, in modernity, have traditionally been represented as hostile and not fit for human habitation. In the case of Tianjin, a salt marsh environment is being transformed into the eco-city. In Masdar eco-city, the ultimate form of negative 'first nature' (the desert) is becoming a low-carbon oasis characterised by breezes and cool urban temperatures.

Both eco-cities are, therefore, based on foundational stories which pit the new city against a prior, negative natural environment.[14] This helps justify, to some extent, the eco-cities' highly techno-centric visions of what a new urban area must be: technology becomes constructed as the means through which a hostile nature is subdued, controlled, and rationalised by the eco-city. At the same time, it is interesting to note that constructions of a hostile 'first nature' are by themselves *not* faithful representations of the ways in which these environments (salt marsh and desert) were perceived before the contemporary era. In the case of Tianjin, the second chapter highlighted the more positive and nuanced view of the role of the marshland in local culture and in the regional economy. With regards to Masdar eco-city, the 'greening' of the desert is constructed as progress.[15] Although oases have deep cultural significance in the area, nonetheless the construction of the desert as a wholly negative environment which is to be ameliorated by the eco-city belies the need to use a 'hostile' natural environment as a contrast to a new, positive urban landscape. In so doing, technology (from solar power to wind towers) becomes the interface through which transformation (of the desert environment into a cooler, greener, and more sustainable landscape for urban life) is achieved.

DOI: 10.1057/9781137298768.0007

Crisis, the market, and the eco-city

The cases of Tianjin and Masdar eco-cities represent some of the most technically advanced eco-urban projects under construction at the time of writing. They are certainly the largest eco-cities currently being built, and are thoroughly deserving of the 'mega-project' title. In vision and scope, and in the involvement of vast amounts of human and financial capital, they are enormously ambitious. Far from being seen just as new cities by the governments of, respectively, China and Abu Dhabi, they are being treated as experimental sites where entrepreneurial and market-facing policies and systems of governance aim to turn these cities into both economically and environmentally sustainable urban environments.

Both of the eco-cities discussed above are fascinating urban projects in their own right. However, as mentioned in Chapter 1, these mega-projects are part and parcel of a broader trend that has seen the recent emergence of eco-urbanism as a planning and policy trend aimed at building 'greener' cities. In part, this trend is built on a foundation of unease around the notion that the city is facing increasingly systemic dangers, or crises. On the other hand, mainstream eco-urbanism has also incorporated the ideal of the city as an engine for economic growth. Thus, eco-urban projects such as Tianjin and Masdar eco-cities carry within them the twin but not necessarily concordant aims of building more ecologically sensitive urban areas while not impacting on the economic (growth-focused) role of the city.

At the same time, the attention to detail that eco-urban projects such as these eco-cities devote to technology, the market, and infrastructure seems to stand in direct contrast to the paucity of planning for social sustainability in new eco-cities. This contrast is all the more striking in light of the concerns over demographic change, migration, and social 'harmony' that are evident in the national contexts where eco-cities are being built. It therefore seems clear that if eco-urbanism is to make progressive changes to the urban arena, there needs to be more consistent engagement both with the implicit notions of crisis that underpin and help to justify many current eco-city projects, and with the potential for social sustainability which stems from a re-orientation of new cities towards the human inhabitants of these future urban environments.

DOI: 10.1057/9781137298768.0007

Questioning the notion of crisis

As highlighted above, one of the broad contexts within which contemporary eco-cities are being built, and within which the 'green economy' is being defined, is that of the Age of Crisis. The notion of crisis lies at the discursive, cultural, and economic heart of many contemporary urban projects. The construction of new built environments is envisioned and marketed as a way of transitioning away from crisis scenarios in the best of cases or, at worst, as a way of delineating 'protected' environments which can 'mitigate' the negative impacts of future crisis. Deeply embedded within notions of crisis is a view of the city that is essentially negative and sterilising: the city has become constructed as the cause and focus of environmental, social, demographic, and other fears and potential crises, and (technologically) radical new-builds are presented as the only ways in which urban life can continue, or evolve sustainably.

At a more basic level, current debates around sustainable cities reflect a deeper apprehension about urban development in modernity. This concern has deep roots. In an interview given in 1935, for example, Carl Jung touched upon a deep sense of unease with urban modernity when he stated that:

> We are awakening to a feeling that something is wrong in the world, that our modern prejudice of overestimating the importance of the intellect and the conscious mind might be false. We want simplicity. We are suffering in our cities, from a need of simple things. We would like to see our great railroad terminals deserted, the streets deserted, a great peace descend upon us.
>
> These things are being expressed in thousands of dreams... The dream is in large part a warning of something to come... When whole countries avoid these warnings, and fill their asylums, become uniformly neurotic, we are in great danger... Our unconscious wish for deserted places, quiet, inactivity, which now and then is being expressed in the heart of our great cities by a lyrical outbreak of some poet or madman, may project us, against our conscious wills, into another catastrophe from which we may never recover.[16]

One wonders whether blueprints for clean, quiet, sanitised, and technologically advanced eco-cities are not yet another iteration of millenarian dreams of the empty, frictionless city. It is in dreams of cities such as these that, perhaps, the messy life of the urban sphere – embodied in its people – is swept aside or, in the case of the poor, excluded. This

DOI: 10.1057/9781137298768.0007

point is also made in *Evangelii Gaudium*, Pope Francis' latest encyclical, in which a whole section is dedicated to urban cultures. Francis critiques the shape of current urban development (in a way which could be readily applied to the eco-cities described in this book), especially in cases where 'Houses and neighbourhoods are more often built to isolate and protect than to connect and integrate' and where 'non-citizens', 'half citizens', and 'urban remnants' are not given adequate attention by policymakers and society at large.[17] The migrants and construction workers who have built Tianjin and Masdar eco-cities fall within these categories, together with long-term residents of cities reshaped and convulsed by the 'strategic' redesigning of the city, by capital inflows, and by subsequent shifts in the urban landscape.

The idea of crisis is thus of key importance in understanding contemporary eco-cities. The apocalyptic visions that drive the constructed need for new urban centres need to be brought to the light and questioned.[18] The use of discourses of crisis has the effect both of emptying the discursive arena around eco-urbanism of the potential for debate, political engagement, or the voicing of alternatives or dissent (because 'crisis' can conveniently be used as a reason to override such concerns), and of displacing the sphere of action away from the 'public' and into the realm of the 'experts' who are seen as most able to engineer solutions to crisis.

The eco-city can be seen as the material accretion of a range of technological and not completely market-based responses to constructed notions of crisis. These projects are not completely market-based because, as seen in the case of Tianjin and Masdar eco-cities, the state is a prominent actor in shaping new visions of eco-urbanism. While crisis is the starting point for many eco-city projects, the target of 'sustainability' and of a 'sustainable' new urban environment is often the ideal end-point. Thus, while notions of 'crisis' have to be excavated and critically interrogated with regards to eco-urbanism, definitions and socio-political constructions of notions of urban sustainability also have to be engaged with. In the case of Tianjin and Masdar eco-cities, it is not clear that the two cities aim for the same type of sustainability, or that the term is understood as anything more than a container of a range of technical indicators selected by expert actors and with little in the way of public participation or engagement with more socially nuanced notions of what a sustainable eco-city could mean. Getting to grips with 'crisis' and 'sustainability' in the eco-city thus becomes important as a way

DOI: 10.1057/9781137298768.0007

of debating how eco-cities should be shaped (from motivations to the design of the desired new urban environment) and ensuring that current iterations of eco-urbanism do not fall short of their potential by falling back on normative, high modernist visions of the ideal city.

The future city as 'green marketplace'

The concern with urban futures is not simply the remit of urban scholars or of the world of technical knowledge and 'solutions,' or of transition theorists, climate change scientists, and international bodies such as the UN. A host of different groups and organisations have become active in expressing anxiety about the urban future. Some of these concerns are clearly environmental, and have become the focus of environmental advocacy groups, NGOs, and other 'green' organisations.

Nonetheless, much attention has been lavished on the economic significance and viability of eco-city projects. In part, this is understandable, since new-build projects involve large amounts of up-front risk and capital expenditure, and they absorb decades (in aggregate, millennia) of human labour. Furthermore, it is also evident that new eco-cities such as Tianjin or Masdar are not simply envisioned as green and pleasant towns, but as economic drivers of transition towards different economic and industrial ways of organising the economy. In this context, a focus on the economic characteristics of these new urban environments is paramount: the history of urban and industrial projects is littered with the carcasses of 'white elephants' and with the ruins of 'cathedrals in the desert'. The level of political and reputational risk issuing from project failure is significant, and partly explains the focus on envisioning eco-cities as *economic* cities.

Nonetheless, some of the overwhelming attention paid to the economic nature and role of eco-cities reflects corporate, technology-focused notions of these cities as a package of technical solutions to 'crisis.' In many cases, these visions are thinly veiled conceptualisations of the future city as a grand marketplace for high-tech products, from sensors for Smart Cities, to hybrid vehicles, to 'smart glass' windows able to save energy as well as display endless advertising.

An example of visions of the 'future city as green marketplace' is the recently opened Crystal exhibition centre near London's Royal Victoria Docks. The centre is a showcase of visions of sustainable cities

DOI: 10.1057/9781137298768.0007

by Siemens. Siemens has developed a corporate focus on 'sustainable cities,' and The Crystal is an exhibition of a range of different 'solutions' aimed at enhancing mobility and quality of life, improving healthcare, and ensuring the viability of Smart Cities. Clearly, the exhibition centre reflects a strong interest in the future of the city, and in what this may mean for technology-focused firms such as Siemens. It is also clear that such techno-centric visions do little to pose important questions around improving cities' socio-economic disparities, inequalities, and injustices. This is of course in great part due to the fact that corporations such as Siemens are not social enterprises: therefore, it is to be expected that The Crystal's exhibition space evidences little focus on cities which are socially as well as techno-environmentally sustainable. And yet, a key question remains as to *where* exactly one must turn to find such a focus.

As shown in the previous chapters, many governments and policy-makers seem to have wholeheartedly embraced technology as the panacea for all urban ills. In the meantime, questions of social and economic justice in the sustainable city are conveniently swept under the carpet. A more balanced view of the eco-city must surely focus on how technology can be used in the city to achieve positive socio-environmental and economic outcomes, not at the expense of citizens, but *for* citizens. As Simon Kuznets, winner of the Nobel Prize for economics, argued when discussing the role of advanced technology in society:

> Advancing technology is the *permissive* source of economic growth, but it is only a potential, a necessary condition, in itself not sufficient. If technology is to be employed efficiently and widely, and, indeed, if its own progress is to be stimulated by such use, institutional and ideological adjustments must be made to effect the proper use of innovations generated by the advancing stock of human knowledge. To cite examples from modern economic growth: steam and electric power and the large-scale plants needed to exploit them are not compatible with family enterprise, illiteracy, or slavery – all of which prevailed in earlier times over much of even the developed world, and had to be replaced by more appropriate institutions and social views.[19]

The question remaining therefore is to what extent eco-cities can be truly innovative in economic and technological terms, if their socio-economic constitution does not challenge (and in some cases potentially promotes and deepens) existing inequalities. It seems to follow that in order to be innovative and transitions-focused, eco-city projects could benefit from a re-orientation away from the city-as-marketplace for new

DOI: 10.1057/9781137298768.0007

technologies, and towards the city as a better socio-environmental and economic environment. The city thus becomes focused on its citizens, and not on serving as a vulnerable receptacle of fickle flows of capital and political goodwill.

Rethinking the eco-city

This book has largely been critical of the eco-city projects it has considered. However, it is also important not to completely write them off; after all, at a time when there is much hand-wringing about climate change and urban sustainability, countries such as China and the UAE are actually busily shaping new urban areas – even if, in many cases, the motivations behind new projects seem to be green in the financial sense only. Thus, the point is that if sustainability is seen as environmental, social, and economic, many leading contemporary eco-city projects are deeply economic, somewhat environmental, and potentially not very socially sustainable. Can anything else be expected when cities are a reflection of the current mode of economic and social organisation? A tempting answer may be to state that if cities are a reflection, and materialisation of the economy, then the best that can be achieved is a tinkering round the edges of the city, so as to ensure that the sharpest edges of the free market do not cause too much damage to urban society.

However, to focus on systemic changes in the macro-geography of ideological, economic, and financial systems as the only mechanism through which radical and lasting urban change can occur, conveniently sidesteps the key question of how *specific* urban areas can be made more socially as well as economically and environmentally sustainable. A strictly structural argument that repeatedly points to the sometimes vague and yet epochal landscape of global capitalism and flows of finance as the predominant sites where pressure has to be exerted in order for change to occur on the ground risks missing out on some key factors involved in urban change. Above all, such a focus also carries with it the risk of non-engagement with the specific geographies of city projects, such as the ones considered in this book.

Furthermore, emphasis on the economic determinants of eco-city development belies the fact that cities are not simply expressions or reflections of economic life, but of a more complex and variegated *human* life. Studies of socio-technical transition, for example, have

DOI: 10.1057/9781137298768.0007

strongly pointed to the fact that, often, technological change in the city is related to *cultural* change in the urban and national context. This is seen in the example of the introduction of piped clean water systems in the Netherlands. Clean water started flowing partly as a consequence of elites developing hygienic sensibilities which sparked the rise of a niche market around hygienic products and lifestyles.[20] Within urban studies, there have likewise been recent calls to re-engage with the notion of a 'good' city (where notions of the 'good' are sensitively and broadly conceived).[21] As Harvard University urban planner Susan Fainstein has argued, planning needs to point to:

> [T]he conscious formulation of goals and means for metropolitan development, regardless of whether these determinations are conducted by people officially designated as planners or not. It begins with the premise that a city should be purposefully shaped rather than the unmediated outcome of the market and of interactions within civil society – in other words that planning is a necessary condition for attaining urban values.[22]

In this light, it is useful to highlight alternatives to the sorts of eco-urban visions embodied in the cases explored in previous chapters. There exist a range of alternative iterations of eco-urbanism that are not given much (if any) space outside the academic literature or reports by specific interest groups. These alternatives include, but are not limited to, examples such as the community-led networks of 'Transition Towns', which focus on localised and resilient economies and which are being adopted in the UK and Europe.[23]

Examples of other alternatives to eco-city mega-projects include Low Impact Developments (LIDs), radical attempts to re-integrate urban life into a more natural footprint – similar, in some respects at least, to the ideals underlying Frank Lloyd Wright's Broadacre City.[24] While most of these alternatives have been developed and tested in European urban and political contexts, a broader consideration and adaptation of alternative and radical urban development models defined outside the Global North is also long overdue. This is, for example, the focus of writer Justin McGuirk's recent call to consider the radical and activist architects who are engaging with the urban potential left behind after the crumbling of the modernist dream in Latin America.[25]

The examples mentioned above are not just substitutes for centrally planned, new-build eco-cities, but alternatives to the sort of eco-urbanism that protects, at its core, a continuation of a specific economic way of doing, with its implicit social injustice. This emphasis does not imply a

DOI: 10.1057/9781137298768.0007

rejection of the economy or of technology *per se*, but a recalibration of the balance between the needs of urban dwellers and the requirements of the 'city' seen as a predominantly economic project. Clearly, technology is central to low-carbon trajectories, and economic progress is key to developing the sorts of innovations which will help deliver cleaner cities more able to exist within reduced ecological footprints. However, many of the radical and progressive alternatives that were mentioned above focus not only on technology, but also on consumption. Consumption regimes are as central to the achievement of low-carbon goals as are the latest green and eco technologies. When considering consumption and environmental behaviour, there is clearly a risk of producing and propagating a highly normative, individualised, and moralistic set of directives and norms for the 'good' eco-city. However, while bearing this caution in mind, consumption patterns, preferences, and choices should clearly be important parts of any project which seeks to design a more sustainable urban environment.

Furthermore, when considering the role of alternatives, technology, consumption, and environmental behaviour, it is crucial to consider those aspects of new-build and radical alternatives to contemporary eco-cities that will help ensure more socially just urban environments. Plans for eco-cities need to incorporate a diversity of elements aimed at achieving high standards of socio-environmental equity and justice. Additionally, but crucially if the phenomenon of 'green gating' is to be avoided, socio-economic diversity needs to be a key component of eco-city plans.

On this last point, moves towards planning for greater social sustainability need to be accompanied by a rethink of the relationship between the city and its citizens. Through technology and different 'ways of doing,' eco-cities attempt to stimulate and promote the development of what could be termed the 'eco-urban citizen'. Nonetheless, one of the failures of eco-city projects such as Tianjin and Masdar (although the same critique can be applied to eco-urban projects more widely as well) is precisely in the understanding – by planners, engineers, governments, and development corporations – of the individuals and households who are to populate the new eco-cities of the future.

As seen in the preceding two chapters, these new eco-urban citizens are most often conceptualised first and foremost as simply *economic* actors, playing a part in the functioning of the eco-city as an entrepreneurial, economic 'machine'. It is this definition of the eco-urban citizen

DOI: 10.1057/9781137298768.0007

that robs new eco-city plans of their human scale. The conception of the individual as a merely economic actor is not, however, derived from a single planning or ideological tradition, as seen in the cases of Tianjin and Masdar eco-cities, which stem from different planning contexts and yet display a similar understanding of the future residents of these new cities. Rather, paraphrasing American futurist Alvin Toffler, the conception of the individual as a mere economic actor, as a 'modular individual' whose worth is transactional and most directly tied to the tasks they can perform, can be found within the broader shift, within late modernity, towards more flexible and fast-paced economic change.[26] A direct result of this broader shift is that cities defined as economic spaces remain empty containers of disembodied flows of capital and technology if the role of their residents is not elaborated.

Thus, in the case of Chinese planning, the failure to properly integrate citizens and households into a more vibrant urban fabric can be seen to stem directly from Mao's socialist vision of urban dwellers as economic entities, based on a Marxist fetishisation of *humanity as labour power*, to be fed into the developmental project of the State. On another level, however, the atomisation of the individual urban citizen, and of his or her role in the wider workings of an urban 'project' conceived of purely in economic-industrial terms, is also deeply ingrained in entrepreneurial, neoliberal, market-based visions of the Chinese eco-city in a context of a national planning system evolving after the demise of pre-reform socialist urban planning. In these visions, urban citizens are both consumers of new, 'better' urban environments, as well as economic actors who will turn the eco-city into a successful economic project, and into an attractive investment vehicle for international capital. In both approaches (socialist urban planning and entrepreneurial, market-based urban development) individual citizens' humanity is reduced to their potential for economic work, and their social value is confined to their role within a development trajectory, defined from above, for the new eco-city.

On a deeper level, the definition of the eco-urban citizen as a mere modular, economic actor leaves open the question of what the current eco-urban trend is leading to in terms of urban futures. There is a stark contrast here with Ebenezer Howard's 19th century plans for Garden Cities. Howard's cities were largely conceived as urban environments that would aid in the liberation of citizens from the polluted and dehumanising urban-industrial environments of the industrial conglomerations of the time. Garden Cities can be critiqued at a number of levels;

DOI: 10.1057/9781137298768.0007

nonetheless, Howard's vision was largely focused on serving the *citizens* of his new cities. The city was thus a mechanism for enabling more socially, environmentally, and economically amenable urban environments to grow and develop.

What can be seen in today's eco-cities is the Garden City ideal turned on its head: cities focused on the market and on technology, where urban dwellers are defined more as attendants working for the urban economic and industrial transition machine rather than as citizens served by the urban blueprints, strategies and technical knowledge of planners, governments and corporations. However, a call for a more citizen-focused vision of the eco-city needs not be a simple re-hashing of notions of progressive and participatory urban planning. This is because calls for urban 'liveability' based on such notions risk reproducing the mechanisms of inequality which impact those societal groups which traditionally have less power and representation (from homemakers to immigrant workers) in the planning process.[27]

In the context of the increasing popularity of, and appetite for, eco-cities, it follows that there is great potential for proposing alternatives to the ways in which urban citizens are conceived of in new eco-urban projects. Think, for example, of the radical potential of reducing the inequalities experienced by the workers who build new eco-cities: the aim of fostering socio-economic sustainability and diversity in the city would surely be served well by offering opportunities for their integration into the cities they had a stake in building. Imagine an eco-city plan in which an eco-apartment is assigned to every worker who participated in its construction: giving each worker a stake in, and ownership of, their own investment in the project in terms of labour power, technical skill, and deep knowledge of the material fabric of the city. Imagine also providing them with the legal means to reside in these new cities, through residence permits and access to urban health care and education for themselves and their families. This would not bankrupt new-build eco-cities, or any city for that matter, but it may create a more harmonious society.

Notes

1   Berkhout, F. (2014) Anthropocene futures. *The Anthropocene Review* 1(2): 154–9.

DOI: 10.1057/9781137298768.0007

2   Pendall, R., Foster, K. A. and Cowell, M. (2009) Resilience and regions: building understanding of the metaphor. *Cambridge Journal of Regions, Economy and Society* 3: 71–84.

3   Caragliu, A., Del Bo, C. and Nijkamp, P. (2011) Smart Cities in Europe. *Journal of Urban Technology* 18(2): 65–82; Carvalho, L. (2014) Smart cities from scratch? A socio-technical perspective. *Cambridge Journal of Regions, Economy and Society* DOI: 10.1093/cjres/rsu010; Crivello, S. (2014) Urban policy mobilities: the case of Turin as a Smart City. *European Planning Studies* DOI:10.1080/09654313.2014.891568.

4   Gibbs, D., Krueger, R. and MacLeod, G. (2013) Grappling with Smart City politics in an era of market triumphalism. *Urban Studies* 50(11): 2151–7.

5   Hollands, R. G. (2008) Will the real smart city please stand up? Intelligent, progressive or entrepreneurial? *City: Analysis of Urban Trends, Culture, Theory, Policy, Action* 12(3): 303–20; Allwinkle, S. and Cruickshank, P. (2011) Creating Smart-er cities: an overview. *Journal of Urban Technology* 18(2): 1–16.

6   Lim, C. J. and Liu, E. (2010) *Smart Cities and Eco-Warriors*. London, Routledge: 7.

7   Bailey, I. and Caprotti, F. (2014) The green economy: functional domains and theoretical directions of enquiry. *Environment and Planning A* 46: 1797–813.

8   Reiche, *Renewable energy policies in the Gulf.*

9   Khirfan and Jaffer, *Sustainable urbanism in Abu Dhabi.*

10 Mohammad and Sidaway, *Spectacular urbanization.*

11 Gulf Labor (2014) *Gulf Labor's Observations and Recommendations After Visiting Saadiyat Island and Related Sites.* Available at: http://gulflabor.org/wp-content/uploads/2014/04/GL_Report_Apr30.pdf Accessed 1 May 2014.

12 See Saadiyat (2014) Saadiyat Accommodation Village. Available at: http://www.saadiyat.ae/en/about/about-tdic/worker-welfare/saadiyat-construction-village2.html Accessed 1 May 2014. Quote from Gulf Labor, *Gulf Labor's Observations,* 2.

13 Lui, A. (2013) The 'Island of Happiness: The Shadowy Story of the Guggenheim Abu Dhabi.' *Uncube Magazine.* 8: 44–51.

14 Hinchliffe, S. (1999) 'Cities and natures: intimate strangers.' In Allen, J., Massey, D., and Pryke, M., ed. (1999) *Unsettling Cities* London, Routledge: 141–85.

15 Ryan, C. and Stewart, M. (2009) Eco-tourism and luxury – the case of Al-Maha, Dubai. *Journal of Sustainable Tourism* 17(3): 287–301.

16 Jung, C. G. (1935) 'Americans must say 'No.'' In Sabini, M., ed. (2008) *C. G. Jung on Nature, Technology & Modern Life* Berkeley, CA: North Atlantic Books, 150–1.

17 Francis (2013) *Evangelii Gaudium.* Vatican City: Vatican Press, 62.

18 Mol and Spaargaren, *Environment, modernity and the risk society;* Swyngedouw, *Apocalypse forever?*

19 Kuznets, S. (1973) Modern economic growth: findings and reflections. *The American Economic Review* 63(3): 247–58.

20 Geels, *Co-evolution of technology and society.*

21 Healey, P. (1997) *Collaborative Planning: Shaping Places in Fragmented Societies.* Vancouver, UBC Press.

22 Fainstein, S. (1999) Can we make the cities we want? In Body-Gendrot, S. and Beauregard, R., (eds) (1999) *The Urban Moment: Cosmopolitan Essays on the Late 20th Century City* Thousand Oaks, CA: Sage, 249–72, 250.

23 Hopkins, R. (2008) *The Transition Handbook: From Oil Dependency to Local Resilience.* Cambridge, Green Books; Taylor, P. J. (2012) Transition towns and world cities: towards green networks of cities. *Local Environment: The International Journal of Justice and Sustainability* 17(4): 495–508.

24 Pickerill, J. and Maxey, L. (2009) Geographies of sustainability: Low Impact Developments and radical spaces of innovation. *Geography Compass* 3(4): 1515–39.

25 McGuirk, J. (2014) *Radical Cities: Across Latin America in Search of a New Architecture* London, Verso.

26 Toffler, A. (1972) *Future Shock.* London, Pan Books.

27 Kataoka, S. (2009) Vancouverism: actualizing the livable city paradox. *Berkeley Planning Journal* 22(1): 42–57.

DOI: 10.1057/9781137298768.0007

Bibliography

Abramson, D. B. (2008) Haussmann and Le Corbusier in China: land control and the design of streets in urban redevelopment. *Journal of Urban Design* 13(2): 231–56.

Acuto, M. (2010) High-rise Dubai urban entrepreneurialism and the technology of symbolic power. *Cities* 27(4): 272–84.

Allwinkle, S. and Cruickshank, P. (2011) Creating Smart-er cities: an overview. *Journal of Urban Technology* 18(2): 1–16.

Almond, G. A., Chodorow, M. and Pearce, R. H. (1982) Historical, ideological, and evolutionary aspects. In Almond, G. A., Chodorow, M. and Pearce, R. H. (eds) *Progress and its Discontents*. Berkeley, University of California Press: 17–20.

Al-Sallal, K. A., Al-Rais, L. and Bin Dalmouk, M. B. (2012) Designing a sustainable house in the desert of Abu Dhabi. *Renewable Energy* 49: 80–4.

Baeumler, A., Chen M., Iuchi, K. and Suzuki, H. (2012) Eco-cities and low carbon cities: the China context and global perspectives. In Baeumler, A., Ijjasz-Vasquez, E. and S. Mehndiratta (eds) *Sustainable Low-Carbon City Development in China*. Washington, DC: The World Bank: 33–62.

Baeumler, A., Ijjasz-Vasquez, E. and Mehndiratta, S. (eds) (2012) *Sustainable Low-Carbon City Development in China*. Washington, DC: The World Bank.

Baeumler, A., Chen, M., Dastur, A., Zhang, Y., Filewood, R., Al-Jamal, K., Peterson, C., Randale, M. and

DOI: 10.1057/9781137298768.0008

Pinnoi, N. (2009) *Sino-Singapore Tianjin Eco-City: A Case Study of an Emerging Eco-City in China*. Washington, DC: The World Bank.

Bailey, I. and Caprotti, F. (2014) The green economy: functional domains and theoretical directions of enquiry. *Environment and Planning A* 46: 1797–813.

Bailey, I. and Wilson, G. (2009) Theorising transitional pathways in response to climate change: technocentrism, ecocentrism, and the carbon economy. *Environment and Planning A* 41(10): 2324–41.

Bauman, Z. (2007) *Liquid Times: Living in an Age of Uncertainty* Cambridge, Polity Press.

Beck, U. (1992) *Risk Society*. London, Sage.

Berkhout, F. (2014) Anthropocene futures. *The Anthropocene Review* 1(2): 154–9.

Berndt, C. and Boeckler, M. (2009) Geographies of circulation and exchange: constructions of markets. *Progress in Human Geography* 33: 535–51.

Bernstein S., Lerner J., and Schoar, A. (2013) The investment strategies of sovereign wealth funds. *Journal of Economic Perspectives* 27(2): 219–38.

Boland, A. and Zhu, J. (2012) Public participation in China's green communities: mobilizing memories and structuring incentives. *Geoforum* 43: 147–57.

Boland, A. (2007) The trickle-down effect: ideology and the development of premium water networks in China's cities. *International Journal of Urban and Regional Research* 31(1): 21–40, 30.

Bräutigam, D. and Tang, X. (2011) African Shenzhen: China's Special Economic Zones in Africa. *Journal of Modern African Studies* 49(1): 27–54.

Bray, D. (2006) Building 'community': new strategies of urban governance in China. *Economy and Society* 35(4): 530–49.

Brey, P. (2003) Theorizing modernity and technology. In: Misa T. J., Brey, P. and Feenberg, A. (eds) *Modernity and Technology* Cambridge, MA, MIT Press: 33–71.

Brooker, D. (2012) "Build it and they will come"? A critical examination of utopian planning practices and their socio-spatial impacts in Malaysia's "intelligent city." *Asian Geographer* 29(1): 39–56.

Brown, H. S. and Vergragt, P. J. (2008) Bounded socio-technical experiments as agents of systemic change: the case of a zero-energy

DOI: 10.1057/9781137298768.0008

residential building. *Technological Forecasting & Social Change* 75: 107–30.

Bulkeley, H. (2013) *Cities and Climate Change*. London, Routledge.

Bulkeley, H. and Castán Broto, V. (2012) Government by experiment? Global cities and the governing of climate change. *Transactions of the Institute of British Geographers* 38: 361–75.

Butt, G. (2001) Oil and gas in the UAE. In Al Abed, I. and Hellyer, P. (eds) *United Arab Emirates: A New Perspective*. London, Trident Press: 231–48.

Caprotti, F. (2014) Critical research on eco-cities? A walk through the Sino-Singapore Tianjin Eco-City. *Cities: The International Journal of Urban Policy and Planning* 36: 10–17.

Caprotti, F. (2012) The cultural economy of cleantech: environmental discourse and the emergence of a new technology sector. *Transactions of the Institute of British Geographers* 37(3): 370–85.

Caprotti, F. and Romanowicz, J. (2013) Thermal eco-cities: green building and urban thermal metabolism. *International Journal of Urban and Regional Research* 37(6): 1949–67.

Caragliu, A., Del Bo, C. and Nijkamp, P. (2011) Smart Cities in Europe. *Journal of Urban Technology* 18(2): 65–82.

Carvalho, L. (2014) Smart cities from scratch? A socio-technical perspective. *Cambridge Journal of Regions, Economy and Society* DOI: 10.1093/cjres/rsu010.

Castán Broto, V. (2012) Social housing and low carbon transitions in Ljubljana, Slovenia. *Environmental Innovation and Societal Transitions* 2: 82–97.

Castán Broto, V., Glendinning, S., Dewberry, E., Walsh, C. and Powell, M. (2014) What can we learn about transitions for sustainability from infrastructure shocks? *Technological Forecasting and Social Change* 84: 186–96.

Castells, M. (2010). *End of Millennium, 2nd Ed*. Oxford, Wiley-Blackwell.

Chang, C. I-C. and Sheppard, E. (2013) China's eco-cities as variegated urban sustainability: Dongtan eco-city and Chongming eco-island. *Journal of Urban Technology* 20(1): 57–75.

Chien, S.-S. (2013) Chinese eco-cities: a perspective of land-speculation-oriented local entrepreneurialism. *China Information* 27: 173–96.

China Daily (2011) China's 2011 average salaries revealed. *China Daily* website, 6 July 2012. Available at: http://www.chinadaily.com.cn/china/2012-07/06/content_15555503.htm Accessed 1 March 2014.

DOI: 10.1057/9781137298768.0008

China Daily (2009) China eyes 20% renewable energy by 2020. *China Daily*, 10 June 2009. Available at: http://www.chinadaily. com.cn/china/2009-06/10/content_8268871.htm Accessed 1 March 2014.

China Labour Bulletin (2014) Migrant worker wages increased by 14 percent in 2013. *China Labour* Bulletin website, 21 February 2014. Available at: http://www.clb.org.hk/en/content/migrant-worker-wages-increased-14-percent-2013 Accessed 3 September.

Chu, A. and Chan, J. (2013) *Cleantech in China: Building a Green Future*. London: PricewaterhouseCoopers.

Coaffee, J. (2009) *Terrorism, Risk and the Global City: Towards Urban Resilience*. Farnham, Ashgate.

Coenen, L., Benneworth, P. and Truffer, B. (2012) Toward a spatial perspective on sustainability transitions. *Research Policy* 41(6): 968–79.

Cortesi, A. (1932) The famous Pontine Marshes turned into smiling fields. *The New York Times*, 30 October 1932, XX4.

Crivello, S. (2014) Urban policy mobilities: the case of Turin as a Smart City. *European Planning Studies* DOI:10.1080/09654313.2014.891568.

Cronon, W. (1991) *Nature's Metropolis: Chicago and the Great West*. London, Norton.

Cugurullo, F. (2013) The business of utopia: Estidama and the road to the sustainable city. *Utopian Studies* 24(1): 66–88.

Cugurullo, F. (2013) How to build a sandcastle: an analysis of the genesis and development of Masdar City. *Journal of Urban Technology* 20(1): 23–37.

Datta, A. (2012) India's ecocity? Environment, urbanisation, and mobility in the making of Lavasa. *Environment and Planning C: Government and Policy*, 30(6): 982–96.

Davidson, C. (2009) *Abu Dhabi: Oil and Beyond*. New York, NY: Columbia University Press.

Davidson, C. (2010) Abu Dhabi's global economy: integration and innovation. *Encounters* 1(2): 101–28.

de Jong, M., Yu, C., Chen, X., Wang, D. and Weijnen, M. (2013) Developing robust organizational frameworks for Sino-foreign eco-cities: comparing Sino-Dutch Shenzhen Low Carbon City with other initiatives. *Journal of Cleaner Production* 57: 209–20.

Dempsey, M. C. (2014) *Castles in the Sand: A City Planner in Abu Dhabi*. Jefferson, NC: McFarland & Company.

DOI: 10.1057/9781137298768.0008

Dodds, K. and Ingram, A. (eds) (2009) *Spaces of Security and Insecurity: Geographies of the War on Terror*. Farnham, Ashgate.

Ecocity Builders website (2014) Ecocity Builders. Available at: http://www.ecocitybuilders.org Accessed 1 March 2014.

Economy, E. C. (2007) The Great Leap Backward? The costs of China's environmental crisis. *Foreign Affairs* 86(5): 38–59.

Elliott, D. (2012) *Fukushima: Impacts and Implications*. Basingstoke, Palgrave Macmillan.

El Mallakh, R. (1970) The challenge of affluence: Abu Dhabi. *Middle East Journal* 24(2): 135–46.

Elsheshtawy, Y. (2008) 'Cities of sand and fog: Abu Dhabi's arrival on the global scene.' In Elsheshtawy, Y. (ed.) *The Evolving Arab City: Tradition, Modernity & Urban Development*. London, Routledge: 258–304.

Engwicht, D. (1992) *Towards an Eco-City: Calming the Traffic*. Sydney, Envirobook.

Engwicht, D. (1993) *Reclaiming Our Cities and Towns: Better Living With Less Traffic*. Gabriola Island, BC, Canada: New Society Publishers.

Fainstein, S. (1999) Can we make the cities we want? In Body-Gendrot, S. and Beauregard, R. (eds) (1999) *The Urban Moment: Cosmopolitan Essays on the Late 20th Century City* Thousand Oaks, CA: Sage, 249–72, 250.

Fang, C., Wang, M. and Yang, D. (2013) Understanding changing trends in Chinese wages. In Huang, Y. and Yu, M. (eds) *China's New Role in the World Economy*. London, Routledge: 69–88.

Foster, J. B. (1997) The crisis of the Earth: Marx's theory of ecological sustainability as a Nature-imposed necessity for human production. *Organization & Environment* 10: 278–95.

Francis (2013) *Evangelii Gaudium*. Vatican City, Vatican Press.

Fukuyama, F. (1992) *The End of History and the Last Man*. New York, NY: Free Press.

Gandy, M. (1999) The Paris sewers and the rationalization of urban space. *Transactions of the Institute of British Geographers* 24(1): 23–44.

Gasparini, P., Manfredi, G. and Asprone, D. (eds) (2014) *Resilience and Sustainability in Relation to Natural Disasters: A Challenge for Future Cities*. New York, Springer.

Geels, F. W. (2002) Technological transitions as evolutionary reconfiguration processes: a multi-level perspective and a case-study. *Research Policy* 31(8–9): 1257–74.

DOI: 10.1057/9781137298768.0008

Geels, F. (2005) Co-evolution of technology and society: the transition in water supply and personal hygiene in the Netherlands (1850–1930) – a case study in multi-level perspective. *Technology in Society* 27: 363–97.

Geels, F.W. and Schot, J.W. (2007) Typology of sociotechnical transition pathways *Research Policy* 36(3): 399–417.

Geels, F. W. and Verhees, B. (2011) Cultural legitimacy and framing struggles in innovation journeys: a cultural-performative perspective and a case study of Dutch nuclear energy (1945–1986). *Technological Forecasting and Social Change* 78(6): 910–30.

Ghazal, R. (2011) When Abu Dhabi had 30 cars. *The National*, 3 November 2011. Available at: http://www.thenational.ae/news/uae-news/heritage/when-abu-dhabi-had-30-cars Accessed 1 March 2014.

Gibbs, D., Krueger, R. and MacLeod, G. (2013) Grappling with Smart City politics in an era of market triumphalism. *Urban Studies* 50(11): 2151–7.

Giddens, A. (1990) *The Consequences of Modernity*. Oxford, Polity Press-Balckwell.

Gransow, B. and Daming, Z. (eds) (2010) *Migrants and Health in Urban China*. Berlin, Lit Verlag.

Gulf Labor (2014) *Gulf Labor's Observations and Recommendations After Visiting Saadiyat Island and Related Sites.* Available at: http://gulflabor.org/wp-content/uploads/2014/04/GL_Report_Apr30.pdf Accessed 1 May 2014.

Hajer, M. (1995) *The Politics of Environmental Discourse: Ecological Modernization and the Policy Process*. Oxford, Oxford University Press.

Harvey, D. (1996) *Justice, Nature and the Geography of Difference*. Bognor Regis, Wiley.

Healey, P. (1997) *Collaborative Planning: Shaping Places in Fragmented Societies*. Vancouver, UBC Press.

Heap, T. (2010) Masdar: Abu Dhabi's carbon-neutral city. *BBC News*, 28 March 2010. Available at: http://news.bbc.co.uk/1/hi/world/middle_east/8586046.stm Accessed 1 May 2014.

Hellström Reimer, M. (2010) Unsettling eco-scapes: aesthetic performances for sustainable futures. *Journal of Landscape Architecture* 5(1): 24–37.

Hellyer, P. (2014) End of a 75-year era of oil-fuelled progress for Abu Dhabi. *The National*, 9 January 2014. Available at: http://www.thenational.ae/business/oil/end-of-a-75-year-era-of-oil-fuelled-progress-for-abu-dhabi#page1 Accessed 1 March 2014.

DOI: 10.1057/9781137298768.0008

Hinchliffe, S. (1999) 'Cities and natures: intimate strangers.' In Allen, J., Massey, D., and Pryke, M., ed. (1999) *Unsettling Cities* London, Routledge: 141–85.

Ho, A. L. (2013) Growing pains for Tianjin Eco-City. *The Straits Times Asia Report*, 13 October 2013. Available at: http://www.straitstimes.com/the-big-story/asia-report/china/story/growing-pains-tianjin-eco-city-20131013 Accessed 1 March 2014.

Hodson, M. and Marvin, S. (2010) Urbanism in the anthropocene: ecological urbanism or premium ecological enclaves? *City* 14(3): 299–313.

Hoffman, M. J. (2011) *Climate Governance at the Crossroads: Experimenting with a Global Response*. Oxford, Oxford University Press.

Hollands, R. G. (2008) Will the real smart city please stand up? Intelligent, progressive or entrepreneurial? *City: Analysis of Urban Trends, Culture, Theory, Policy, Action* 12(3): 303–20.

Hopkins, R. (2008) *The Transition Handbook: From Oil Dependency to Local Resilience*. Cambridge, Green Books.

Huang, Y. and Low, S. M. (2008) 'Is gating always exclusionary? A comparative analysis of gated communities in American and Chinese cities.' In Logan, J. R. (ed.) *Urban China in Transition*. Oxford, Blackwell: 182–202.

Isles, M. (2014) Melbourne's eco city transition plan. Local Government Managers Australia, National Office website. Available at: http://www.lgma.org.au/default/melbournes_eco_city_transition_plan Accessed 19 August 2014.

Jackson, M. S., and Della Dora, V. (2011) From landscaping to terraforming? Gulf mega-projects, cartographic visions and urban imaginaries. In: Agnew, J., Roca, Z., and P. Claval (eds) *Landscapes, Identities and Development*. Farnham, Ashgate: 95–113.

Joss, S. and Molella, A. (2013) The eco-city as urban technology: perspectives on Tangshan Caofeidian International Eco-City (China). *Journal of Urban Technology* 20(1): 115–37.

Joss, S., Tomozeiu, D. and Cowley, R. (2011) *Eco-Cities: A Global Survey 2011*. London, University of Westminster.

Joss, S., Tomozeiu, D. and Cowley, R. (2012) Eco-city indicators: governance challenges. In Pacetti, M., Passerini, G., Brebbia, C. A. and Latini, G. (eds) *Sustainable City VII: Urban Regeneration and Sustainability*. Southampton, WIT Press: 109–20.

Jung, C. G. (1935) 'Americans must say 'No.'' In M. Sabini (ed. 2008) *C. G. Jung on Nature, Technology & Modern Life* Berkeley, CA: North Atlantic Books, 150–1.

DOI: 10.1057/9781137298768.0008

Kaika, M. (2010) Architecture and crisis: re-inventing the icon, re-imag(in)ing London and re-branding the City. *Transactions of the Institute of British Geographers* 35(4): 453–74.

Kaika, M. (2006) Dams as symbols of modernisation: the urbanisation of nature between materiality and geographical representation. *Annals of the Association of American Geographers* 96(2): 276–301.

Kaiman, J. (2013) Chinese struggle through 'airpocalypse' smog. *The Guardian*, 16 February 2013. Available at: http://www.theguardian.com/world/2013/feb/16/chinese-struggle-through-airpocalypse-smog Accessed 1 March 2014.

Kargon, R. H. and Molella, A. P. (2008) *Invented Edens: Techno-Cities of the Twentieth Century*. Cambridge, MA: MIT Press.

Kataoka, S. (2009) Vancouverism: actualizing the livable city paradox. *Berkeley Planning Journal* 22(1): 42–57.

Keppel Land (2013) *Across Borders*. 4th quarter, 2013. Keppel Corporation, Singapore.

Khirfan, L. and Jaffer, Z. (2013) Sustainable urbanism in Abu Dhabi: transferring the Vancouver model. *Urban Affairs*. Available at: http://onlinelibrary.wiley.com/doi/10.1111/juaf.12050/full (Accessed 1 March 2014).

Kim, C. (2010) Place promotion and symbolic characterization of New Songdo City, South Korea. *Cities* 27(1): 13–19.

Kuznets, S. (1973) Modern economic growth: findings and reflections. *The American Economic Review* 63(3): 247–58.

Leichenko, R. (2011) Climate change and urban resilience. *Current Opinion in Environmental Sustainability* 3(3): 164–8.

Leung, P. (2012) Wenzhou's once-hot housing market comes crashing down. *South China Morning Post*, 23 December 2012. Available at: http://www.scmp.com/news/china/article/1111003/wenzhous-once-hot-housing-market-has-come-crashing-down Accessed 1 March 2014.

Li, B. (2008) 'Why do migrant workers not participate in urban social security schemes? The case of the construction and service sectors in Tianjin.' In Nilsen, I. and Smyth, R. (eds) *Migration and Social Protection in China*. London, World Scientific: 184–204.

Li, Y. (2012) Environmental state in transformation: the emergence of low-carbon development in urban China. In Holt, W. G. (ed.) *Urban Areas and Global Climate Change*. Bingley, Emerald: 221–46.

Li, Y. and Currie, J. (2011) *Green Buildings in China: Conception, Codes and Certification*. Washington, DC: Johnson Controls.

DOI: 10.1057/9781137298768.0008

Lim, C. J. and Liu, E. (2010) *Smart Cities and Eco-Warriors*. London, Routledge.

Lin, L., Liu, Y., Chen, J., Zhang, T. and Zeng, S. (2011) Comparative analysis of environmental carrying capacity of the Bohai Sea Rim area in China. *Journal of Environmental Monitoring* 13(11): 3178–84.

Liu, C. (2011) China's city of the future rises on a wasteland. *The New York Times*, 28 September 2011. Available at: http://www.nytimes.com/cwire/2011/09/28/28climatewire-chinas-city-of-the-future-rises-on-a-wastela-76934.html?pagewanted=all Accessed 1 March 2014.

Liu, J. and Diamond, J. (2005) China's environment in a globalizing world. *Nature* 435(30): 1179–86.

Liu, W., Lund, H., Mathiesen, B. V. and Zhang, X. (2011) Potential of renewable energy systems in China. *Applied Energy* 88(2): 518–25.

Logan, J. R., ed. (2008) *Urban China in Transition*. Oxford, Blackwell: 182–202.

Lora-Wainwright, A. (2013) *Fighting for Breath: Living Morally and Dying of Cancer in a Chinese Village*. Honolulu, University of Hawai'i Press.

Lora-Wainwright, A. (2010) An anthropology of 'cancer villages': villagers' perspectives and the politics of responsibility. *Journal of Contemporary China* 19(63): 79–99.

Lui, A. (2013) The 'Island of Happiness: The Shadowy Story of the Guggenheim Abu Dhabi.' *Uncube Magazine*. 8: 44–51.

Maalouf, A. (2012) *Disordered World: A Vision for the Post-9/11 World*. London, Bloomsbury Paperbacks.

Mahroum, S. and Alsaleh, Y. (2012) Place branding and place surrogacy: the making of the Masdar cluster in Abu Dhabi. Faculty & Research Working Paper 2012/130/IIPI. INSEAD Abu Dhabi Campus: Abu Dhabi.

Masdar (2007) Abu Dhabi heats up the global solar market with $2 billion investment in photovoltaic manufacturing. Press release, 27 May 2007. Available at: http://www.masdar.ae/en/mediacenter/newsDesc.aspx?News_ID=85&MenuID=55&CatID=44 Accessed 1 May 2014.

Masdar (2008) Abu Dhabi commits US$15 billion to alternative energy, clean technology. Press release, 21 January 2008. Available at: http://www.masdar.ae/en/mediacenter/newsDesc.aspx?News_ID=42&MenuID=55&CatID=44 Accessed 1 May 2014.

Masdar (2008) Abu Dhabi's Masdar Initiative breaks ground on carbon-neutral city of the future. Press release, 9 February 2008.

DOI: 10.1057/9781137298768.0008

Available at: http://www.masdar.ae/en/mediacenter/newsDesc. aspx?News_ID=40&MenuID=55&CatID=44 Accessed 1 May 2014.

Masdar (2010) Masdar partners with total and Abengoa Solar. Press release, 20 June 2010. Available at: http://www.masdar.ae/en/ mediaCenter/newsDesc.aspx?News_ID=144&MenuID=0&CatID=0 Accessed 1 May 2014.

Masdar City (2010) Business. Available at: http://www.masdarcity.ae/en/ index Accessed 1 May 2014.

Masdar Clean Tech Funds (2010). Home: Masdar Clean Tech Funds Available at: http://www.masdarctf.com Accessed 1 May 2014.

Masdar website (2014) Masdar. Available at: http://www.masdar.ae Accessed 1 May 2014.

Matsumoto, T. (2013) Abu Dhabi energy policy: energy problems plaguing Abu Dhabi and their implications for Japan. *Institute of Energy Economics of Japan*, August 2013. Available at: http://eneken. ieej.or.jp/data/5115.pdf Accessed 1 March 2014.

McGuirk, J. (2014) *Radical Cities: Across Latin America in Search of a New Architecture* London, Verso.

Mehzer, T. Dawelbait, G. and Abbas, Z. (2012) Renewable energy policy options for Abu Dhabi: drivers and barriers. *Energy Policy* 42: 315–28.

Meinhold, B. (2009) Masdar breaks ground on largest solar plant in Middle East. *Inhabitat*, 21 January 2009. Available at: http://www. inhabitat.com/2009/01/21/masdar-begins-construction-on-10mw-solar-power-plant Accessed 1 May 2014.

Milcent, C. (2010) Healthcare for migrants in urban China: a new frontier. *China Perspectives* 2010(4): 33–46.

Ministère du Logement et de l'Égalité des Territoires (2014) Les EcoCités. Available at: http://www.territoires.gouv.fr/les-ecocites Accessed 19 August 2014.

Ministry of Finance and the State Administration of Taxation (2006) Incentives of the corporate income tax to support the development of the TBNA. 130(2006), 15 November 2006. Available at: http://en.investteda. org/download/revised20070115.doc Accessed 1 March 2014.

Misa T. J., Brey, P. and Feenberg, A. (eds) *Modernity and Technology*. Cambridge, MA, MIT Press.

Mitchell, T. (2002) *Rule of Experts: Egypt, Techno-Politics, Modernity*. Berkeley, CA: University of California Press.

Mohammad, R. and Sidaway, J. (2013) Spectacular urbanization amidst variegated geographies of globalization: learning from Abu Dhabi's

DOI: 10.1057/9781137298768.0008

trajectory through the lives of South Asian men. *International Journal of Urban and Regional Research* 36: 606–27.

Mol, A.P.J. and Spaargaren, G. (2009) Environment, modernity and the risk-society: the apocalyptic horizon of environmental reform. *International Sociology* 4: 431–59.

Monstadt, J. (2009) Conceptualizing the political ecology of urban infrastructures: insights from technology and urban studies. *Environment and Planning A* 41(8): 1924–42.

MDC (2010) About Mubadala. Available at: http://www.mubadala.ae/en/category/about-mubadala Accessed 1 May 2014.

Murray, M. (2013) Connecting wealth and growth through visionary planning: the case of Abu Dhabi 2030. *Planning Theory & Practice* 14(2): 278–82.

Na, L. (2013) Top 10 most polluted Chinese cities in 2012. *China.org.cn*, 15 April 2013. Available at: http://www.china.org.cn/top10/2013-04/15/content_28541619.htm Accessed 1 March 2014.

Nader, S. (2009) Paths to a low-carbon economy – the Masdar example. *Energy Procedia* 1.1, 3951–8.

Newman, P., Beatley, T. and Boyer, H. (2009) *Resilient Cities: Responding to Peak Oil and Climate Change*. Washington, DC: The Island Press.

Nye, D. E. (1998) *Consuming Power: A Social History of American Energies*. Cambridge, MA: MIT Press.

Oliver, S. (2000) The Thames Embankment and the disciplining of nature in modernity. *The Geographical Journal* 166(3): 227–38.

Ouis, P. (2011) 'And an island never cries': cultural and societal perspectives on the mega development of islands in the United Arab Emirates. In Badescu, V. and Cathcart, R. V. (eds) *Macro-Engineering Seawater in Unique Environments: Arid Lowlands and Water Bodies Rehabilitation*. Berlin, Springer: 60–75, 60.

Oxford Business Group (2010) *The Report: Abu Dhabi 2010*. London, Oxford Business Group.

Pendall, R., Foster, K. A. and Cowell, M. (2009) Resilience and regions: building understanding of the metaphor. *Cambridge Journal of Regions, Economy and Society* 3: 71–84.

Pickerill, J. and Maxey, L. (2009) Geographies of sustainability: Low Impact Developments and radical spaces of innovation. *Geography Compass* 3(4): 1515–39.

Platt, H. L. (1991) *The Electric City: Energy and the Growth of the Chicago Area, 1880–1930*. Chicago, The University of Chicago Press.

DOI: 10.1057/9781137298768.0008

Ponzini, D. (2011) Large scale development projects and star architecture in the absence of democratic politics: the case of Abu Dhabi, UAE. *Cities* 28(3): 251–9.

Pow, C.-P. (2009) *Gated Communities in China: Class, Privilege and the Moral Politics of the Good Life*. London, Routledge.

Rapoport, E. (2014) Utopian visions and real estate dreams: the eco-city past, present and future. *Geography Compass* 8(2): 137–49.

Reed, T. V. (2014). *Digitized Lives: Culture, Power and Social Change in the Internet Era*. London, Routledge.

Register, R. (1987) *Ecocity Berkeley: Building Cities for a Healthy Future*. Berkeley, CA: North Atlantic Books.

Reiche, D. (2010) Renewable energy policies in the Gulf countries: a case study of the carbon-neutral "Masdar City" in Abu Dhabi. *Energy Policy* 38(1): 378–82.

Ren, X. (2013) *Urban China*. Cambridge, Polity Press.

Roelofs, J. (2000) Eco-cities and red green politics. *Capitalism Nature Socialism* 11: 139–48.

Romano, G. (2013) No administrative solution in sight for urban 'Airpocalypse'. *China Perspectives* 3: 82–4.

Roseland, M. (1997) Dimensions of the eco-city. *Cities* 14: 197–202.

Royal Ten Cate (2013) *Ten Cate Annual Report*. Almelo, Royal Ten Cate.

Royal Ten Cate (2013) Ten Cate Geotube® wastewater impoundment lake remediation – Tianjin Eco-City, China. Available at: http://www.tencate.com/apac/geosynthetics/case-studies/dewater-tech/news-dewat3.aspx Accessed 1 March 2014.

Ruano, M. (1999) *Eco-Urbanism: Sustainable Urban Settlements*. Barcelona, Gustavo Gili.

Ryan, C. and Stewart, M. (2009) Eco-tourism and luxury – the case of Al-Maha, Dubai. *Journal of Sustainable Tourism* 17(3): 287–301.

Saadiyat (2014) Saadiyat Accommodation Village. Available at: http://www.saadiyat.ae/en/about/about-tdic/worker-welfare/saadiyat-construction-village2.html Accessed 1 May 2014.

Sabini, M., ed. (2008) *C. G. Jung on Nature, Technology & Modern Life* Berkeley, CA: North Atlantic Books.

Shan, J. and Qian, Y. (2009) Abortion statistics cause for concern. *China Daily*, 30 July 2009. Available at: http://www.chinadaily.com.cn/china/2009-07/30/content_8489656.htm Accessed 1 March 2014.

Shwayri, S. (2013) A model Korean ubiquitous eco-city? The politics of making Songdo. *Journal of Urban Technology* 20(1): 39–55.

DOI: 10.1057/9781137298768.0008

Sim, L-C. (2012) Re-branding Abu Dhabi: from oil giant to energy titan. *Place Branding and Public Diplomacy* 8: 83–98.

Smil, V. (1993) *China's Environmental Crisis: An Enquiry into the Limits of National Development.* New York, M. E. Sharpe.

Sovereign Wealth Fund Institute (2014) Mubadala Development Company Available at: http://www.swfinstitute.org/fund/mubadala.php Accessed 1 May 2014.

Spaargaren, G. and Mol, A. P. J. (1992) Sociology, environment, and modernity: ecological modernization as a theory of social change. *Society & Natural Resources: An International Journal* 5(4): 323–44.

Springer, C. (2012) Public housing in the Sino-Singapore Tianjin Eco-City: the (missing) link between social cohesion and green living. *EcoCity Notes*, May 2012. Available at: http://ecocitynotes.com/features/eco-city-demographics/public-housing-in-the-sino-singapore-tianjin-eco-city/ Accessed 1 March 2014.

Statistics Center Abu Dhabi (2010) *Abu Dhabi Over Half a Century.* Abu Dhabi, SCAD.

SSTECIDC. 2010. Celebrating eco. Available at: http://events.cleantech.com/tianjin/sites/default/files/SSTECBrochureFinal.pdf Accessed 1 March 2014.

Suzuki, H., Dastur, A., Moffatt, S., Yabuki, N. and Maruyama, H. (2010) *Eco2 Cities: Ecological Cities as Economic Cities.* Washington, DC: The World Bank.

Swyngedouw, E. (2010) Apocalypse forever? Post-political populism and the spectre of climate change. *Theory, Culture and Society* 27: 213–32.

Swyngedouw, E. (1999). Modernity and hybridity: nature, *Regeracionismo* and the production of the Spanish waterscape, 1890–1930. *Annals of the Association of American Geographers* 89(3): 443–65.

Taylor, P. J. (2012) Transition towns and world cities: towards green networks of cities. *Local Environment: The International Journal of Justice and Sustainability* 17(4): 495–508.

Teather, D. (2010) Bailed out and broke, Dubai opens the world's tallest building. *The Guardian*, 3 January 2010. Available at: http://www.theguardian.com/business/2010/jan/03/burj-dubai-worlds-tallest-building Accessed 1 March 2014.

TEDA (2010) Tianjin Binhai New Area. Available at: http://en.investteda.org/BinhaiNewArea/default.htm Last accessed 1 March 2014.

DOI: 10.1057/9781137298768.0008

The Economist (2013) Sending the foreigners home. *The Economist*, 13 July 2013. Available at: http://www.economist.com/news/middle-east-and-africa/21581783-sacking-foreign-civil-servants-may-become-regional-trend-sending Accessed 1 March 2014.

Toffler, A. (1972) *Future Shock*. London, Pan Books.

UAEinteract (2013) Abu Dhabi's population at 2.33M, with 475,000 Emiratis. UAEinteract, 9 October 2013. Available at: http://www. uaeinteract.com/docs/Abu_Dhabi's_population_at_2.33m,_ with_475,000_Emiratis/57590.htm Accessed 1 March 2014.

UPC (2007) *Plan Abu Dhabi 2030*. Abu Dhabi: Abu Dhabi Urban Planning Council.

Van Berkel, R., Fujita, T., Hashimoto, S. and Geng, Y. (2009) Industrial and urban symbiosis in Japan: analysis of the Eco-Town program 1997–2006. *Journal of Environmental Management* 90(3): 1544–56.

van Dijk, M. P. (2011) Three ecological cities, examples of different approaches in Asia and Europe. In Wong, T.-C. and Yuen, B. (eds) *Eco-City Planning: Policies, Practice and Design*. New York, Springer: 31–50.

Watson, D. and Adams, M. (2011) *Design for Flooding: Architecture, Landscape, and Urban Design for Resilience to Climate Change*. Hoboken, NJ: John Wiley & Sons.

Wang, F. (2010) China's population destiny: the looming crisis. The Brookings Institution, September 2010. Available at: http://www. brookings.edu/research/articles/2010/09/china-population-wang Accessed 1 March 2014.

Wilson, J. and Anielski, M. (2005) *Ecological Footprints of Canadian Municipalities and Regions*. Edmonton, Federation of Canadian Municipalities.

Wong, T. C. (2011) 'Eco-cities in China: pearls in the sea of degrading urban environments.' In Wong, T.-C. and Yuen, B. (eds) *Eco-City Planning: Policies, Practice and Design*. New York, Springer: 131–50.

Wong, T.-C. and Yuen, B. (eds) (2011) *Eco-City Planning: Policies, Practice and Design*. New York, Springer.

World Future Energy Summit (2012) Exhibition & Summit 2012. Available at: http://www.worldfutureenergysummit.com/Portal/ about-wfes/overview/2011-summit-and-exhibition.aspx Accessed 1 March 2014.

Wu, F. (2013) China's eco-cities. *Geoforum* 43(2): 169–71.

DOI: 10.1057/9781137298768.0008

Yeung, Y.-M., Lee, J. and Kee, G. (2009) China's Special Economic Zones at 30. *Eurasian Geography and Economics* 50(2): 222–40.

Zhang, P. and Xu, M. (2011) The view from the county: China's regional inequalities of socio-economic development. *Annals of Economics and Finance* 12(1): 183–98.

Zhou, S., Dai, J. and Bu, J. (2013) City size distributions in China 1949 to 2010 and the impacts of government policies. *Cities* http://dx.doi.org/10.1016/j.cities.2013.04.011.

DOI: 10.1057/9781137298768.0008

Index

DOI: 10.1057/9781137298768.0009

DOI: 10.1057/9781137298768.0009